Subversion, Sexuality and the Virtual Self

Subversion, Sexuality and the Virtual Self

Jude Elund
Edith Cowan University, Australia

palgrave
macmillan

First published 2015 by
PALGRAVE MACMILLAN

Palgrave Macmillan in the UK is an imprint of Macmillan Publishers Limited, registered in England, company number 785998, of Houndmills, Basingstoke, Hampshire RG21 6XS.

Palgrave Macmillan in the US is a division of St Martin's Press LLC, 175 Fifth Avenue, New York, NY 10010.

Palgrave Macmillan is a global academic imprint of the above companies and has companies and representatives throughout the world.

Palgrave® and Macmillan® are registered trademarks in the United States, the United Kingdom, Europe and other countries.

ISBN 978–1–137–46833–8

This book is printed on paper suitable for recycling and made from fully managed and sustained forest sources. Logging, pulping and manufacturing processes are expected to conform to the environmental regulations of the country of origin.

A catalogue record for this book is available from the British Library.

Library of Congress Cataloging-in-Publication Data
Elund, Jude, 1978–
Subversion, sexuality and the virtual self / Jude Elund, Edith Cowan University, Australia.
pages cm
Summary: "This book is about the construction of identity in digitised and virtual screen environments. It investigates the phenomenon of three-dimensional virtual worlds in relation to our physical selves. Using four regions of Second Life as a foundation of analysis, this book explores the issues surrounding the body, embodiment, virtual space and tourism, as well as examining notions of subversion as they relate to gender and sexuality in screen environments. The key arguments put forth in this work have their foundations in poststructuralist theories of the body and the cultural construction of identity. Most important to this analysis are the cultural and social conditions that replicate dominant tropes of power and agency as they are applied to representational practices, such as embodying oneself via an avatar through the screen. The arguments in this book are readily extendible to other on-screen environments due to the proliferation of virtual environments and other social networks which are premised on real-time engagement through digitized interaction"— Provided by publisher.
ISBN 978–1–137–46833–8 (hardback)
1. Avatars (Virtual reality) 2. Online identities. 3. Shared virtual environments. 4. Virtual reality. 5. Human-computer interaction. 6. Sex role in mass media. 7. Technological innovations—Social aspects. I. Title.
QA76.76.I59E48 2015
006.8—dc23 2015001282

Typeset by MPS Limited, Chennai, India.

Contents

List of Figures

Acknowledgements

The four years researching and crafting this work have been a very rewarding period of my life. It is such a privilege to be able to devote a concentrated time and effort into a project which is incredibly self-satisfying. A good amount of this time was spent in the world of Second Life, so initial thanks should go to the people who devote their time, money and effort in developing the wonderful regions of the platform, some of which are represented in the images within this book. There have been a number of individuals who have provided inspiration and support for the work itself as well as support for, and belief in, me – I am eternally grateful to them all.

Foremost, my supervisor and mentor Associate Professor Panizza Allmark, who I cannot thank enough. Her commitment, support and unwavering belief in my abilities have been the most important factor in my achievements – there really are few words to express my gratitude. Similarly, Professor Jon Stratton has had a huge influence on the early stages of my career – his guidance has been enormously valuable.

I have a huge appreciation for the staff at the School of Communication and Arts, Edith Cowan University and the MCCA School at Curtin University. Specifically, I would like to thank Dr Dennis Wood for introducing the world of theory to me as a young undergraduate, Dr Debbie Rodan for her enthusiasm and mentorship, Bethany Andersson for proof-reading and suggestions and Professor Lelia Green for her support and guidance from my undergraduate years through to post-graduate mentoring, advice and support. I would also like to thank my colleague in everything cultural studies, James Hall, for many stimulating and rewarding discussions.

On a more personal level, I would like to express my thanks and gratitude to my family, especially my father for his interest in my work, and to my dearest friend Dean for his intellect and love.

Introduction

This book is about the construction of identity in digitised and virtual screen environments. It investigates the phenomenon of three-dimensional virtual worlds (3DVEs) in relation to our physical, or corporeal, selves. Using four regions of Second Life as a foundation of analysis, this book explores the issues surrounding the body, embodiment, virtual space and tourism, as well as examining notions of subversion as they relate to gender and sexuality in screen environments. The key arguments put forth in this work have their foundations in poststructuralist theories of the body and the cultural construction of identity, which follow a thematic application of discourse analysis within each chapter as a basis for critiquing Western screen culture. Most important to this analysis are the cultural and social conditions that replicate dominant paradigms of power and agency as they are applied to representational practices, such as embodying oneself via an avatar through the screen. Although mainly focused on Second Life, the arguments in this book are readily extendable to other on-screen environments due to the proliferation of 3DVEs and other social networks which are premised on real-time engagement through digitised and representational interaction.

The potential to embody anything or any body is a phenomenon that is available to many through online platforms such as Second Life (SL). The opportunities provided by three-dimensional virtual environments (3DVEs) are part of their attractiveness and popularity – seemingly they promise escape from our corporeal present into a world of possibility where we can be anything we want. Yet, on entering the world of SL it is immediately apparent that corporeal commonalities and norms of representation exist pervasively throughout its islands, spaces and regions. Whilst we may be able to escape the everyday in a temporal and spatial sense through the screen into a 'second life', this virtual experience remains grounded in our cultural norms and expectations of what a potentially utopian world would look like. Central to this ideation is mass culture, in its representations, values and norms which work ideologically to speak to us about desires, success and happiness. What we see, hear and read in our corporeal lives as exciting, fun and desirable is re-presented in the digital spaces of SL. As is typical of the internet, sexual freedom is a given, yet representational performance (how one appears through ones' embodied avatar) is predicated on significations derived from the corporeal world. Therefore, within a potentially free space, there is a persistent normativity which reiterates the historically and culturally ascribed foundations of identity. This normativity is particularly prevalent in gendered representation – avatars tend to hyper-gendered expression and the excesses of Western bodily presentation and adornment (although virtual bodies exceed corporeal and biological capacity of attainment). Like the idealised corporeal body of mass culture, the virtual body is pushed to the limits of gendered excess and sexual desirability as the hyper-masculine or hyper-feminine subject. That these representational practices carry over into sexually diverse regions (such as the SL regions analysed in this text) is perhaps unsurprising, given that gay and lesbian culture has been in a large way subsumed into contemporary mass culture. It is the tensions that occur as a result of the normative acting upon queer, gay and lesbian and subversive cultures that form the basis of investigation, specifically the relationship between corporeal normativity and screen culture as well as the tensions between cultural conservatism, subversive representation and gender conformity.

The discussion in this book has its foundations in the Western corporeal existence, most importantly the cultural and social conditions that produce and replicate dominant paradigms of power and agency. Of particular pertinence to this discussion are the framing, representational and spatial practices of gendered and sexual identities within SL spaces. There has been a tradition of envisaging future technologies as potentially utopian, where individuals may be freed from their biological, gendered and sexed determinism. According to Donna Haraway (1991, p. 163), to live a life in critique of the corporeal can enact a metaphorical vision of a cyborg future:

> The cyborg is not subject to Foucault's biopolitics; the cyborg simulates politics, a much more potent field of operations ... The cyborg is a kind of disassembled and reassembled, postmodern collective and personal self.

However, the desired emancipatory shift that some theorists had hoped for, in view of the body as political, has not eventuated. This is despite the mass appropriation of technological devices which in many ways free us from our physical selves, in addition to the continued development of devices that allow for ever greater degrees of immersion and presence in an extension of our fixed corporeal reality. Paradoxical to this notion of the self as unbounded by the norms of society, virtual spaces abound with gendered and sexed bodies that are generally highly polarised in terms of gender as well as highly homogenised in terms of what constitutes sexual desirability (as seen in the rise and predominance of "raunch culture" (Levy 2005), with its highly overt sexual displays). Moreover, whilst there has been a shift towards the eroticisation of the digitally represented self, there has been a concurrent conservative shift from radical subjectivities to conformist and normative identities, even in groups that have previously been associated with subversion and difference, such as the lesbian, gay, bisexual, transgender, intersex (LGBTI) community. This modality of visual culture is explored through the representational aspects of identity, in analysing the body as object, body as conformative, body as performative, and body as subversive.

Given the disembodied potential of SL, and the ability to re-embody oneself in a multitude of forms and on-screen identities, the platform allows for extended analysis of the performative and fluid aspects of gender beyond the corporeal. The notion of performativity here rests on Judith Butler's work on gender as performative, where 'acts, gestures, enactments, generally construed, are *performative* in the sense that the essence or identity that they otherwise purport to express are *fabrications* manufactured and sustained through corporeal signs and other discursive means' (1999, p. 173). In the SL environment, the naturalness of the body to its physical enactments, or bodily significations to internal processes such as maleness to masculine performance, are entirely removed; the supposition is revealed as a mirage. The reliance on gender as a symbol of value and exchange within SL is shown as both a commodity and a foundation for identity. Although Butler denounces identity categories as predicated on the false assumption that individuals of a collective sex share intrinsic characteristics, the discussion within this book often departs from hers by conceptualising some gender and sexuality categorisations as useful in the political context (such as within marginalised groups that can find some form of sanctuary and agency together in collectives). However, this is not to essentialise the positions of female, male, lesbian, homosexual et al. as possessing some innate forms of coherence or shared essence. Furthermore, and following from Judith Halberstam (1998), such categorisations are seen as an extension of historical and cultural narratives that cannot be simply removed from notions of selfhood and identity. Rather, identity needs to be viewed as an extension of the self in the historical context, whereby identity markers are appropriated so as to participate as part of a society which has shared meanings about self-representation and signification practices. Gay and lesbian, as well as other non-conformist, identities can therefore be seen to operate within this context, as an extension of modernist categorisations that have mostly necessitated conformity to gender and sexual binaries.

Whilst it is arguable that gendered identity categories are necessary, there are problems when such categorisations become essentialised as extensions of the heterosexual matrix. Within gay culture, this essentialism is evident in the shift towards conservatism and normativity, which may be attributed to a desire for acceptance in the wider community. This shift has led to a further exoticisation and othering of sexual identities that operate outside of the conservative structure, so that even in the already marginalised gay/queer community there are individuals who are now even more marginalised by virtue of a shift towards a homo-normativity. This shift has been remarked upon by such commentators as Michael Warner, who uses the term 'selective legitimacy' (1999, p. 82) to describe such exclusion. Similarly, Paul Robinson (2005) describes the political move towards conservatism of some gay factions having the effect of de-legitimising many behaviours, as well as many non-normative identities. This book investigates the notions of cultural conservatism, including hetero- and homo-normativity, in opposition to the subversive potentialities of SL. Of particular importance is the exploration of corporeal normativity as reflected and performed through the screen, including notions of coupledom and promiscuity, pornographic representation and the gaze, and gender conformity.

Identity in the SL environment is expressed primarily in visual forms of signification and representation. Therefore, bodies within SL can be considered as unfixed from any biological "truth;" they are entirely signified and hyper-real extensions of an individual. Through the transformative potential of creating an avatar, it is impossible to reconcile avatar representation on the screen with any truth of an individual's actual selfhood and/or experiences: an individual's geography, history, ethnicity, age, gender and other bodily markers are suspended in-world. This is perhaps one of the greatest attractions of the environment, as of the generally: anonymity and the liberty to present oneself and interact as one chooses. Similarly, SL spaces and regions have political and cultural relevance dependent on sign values as constructed in corporeal life. As recognised by Lefebvre (1991), space itself is a signifying practice that holds discursive value and can modulate the behaviour of individuals within a given environment. The configuration and conception of space holds within it the power, ideologies and subjective imaginings of those who create it and those who prescribe its use. In this study, both identity and space are utilised as signifying practices for the purpose of analysing power structures. Given that both are user-generated in the SL environment, an analysis of signifying practices shows that there is a pronounced tendency to replicate many of the discursive practices of corporeal life, even in the face of endless possibilities of creation and appropriation.

The freedom of choice associated with embodiment and space within SL offers escapism as well as an active participation in a world that can be seen to exist on the boundary of corporeal and virtual lives. In addition to a possible desired self as signified by avatar selection and adornment, there is the promise of desired (often utopian) spaces which evoke an imagined or idealised realm. This imagined space is discussed in two key ways: firstly, as framed within tourist discourse and the associated desire for hedonism; and secondly, as part of the cultural and historical narratives associated with the Classical world as imagined utopias of an idealised past. Framing the regions in this way allows a reading of the spaces that suggests either a liminal or liminoid appropriation, where the liminal can be seen as a projected engagement with the community and ideals of a certain space, and liminoid is taken as an expression of bodily pleasures as the primary outcome for visiting that space. These modes align with MacCannell's (1999) dyad of authentic/inauthentic experience in reference to the usage patterns and motivations for visiting certain spaces, particularly as they relate to ideas of a mythologised past. In particular, these spaces provide an opportunity for a tourist-like experience in which individuals explore gender and sexuality in an environment that is largely premised on an idealised conception of sexuality and freedom.

Virtual experience

Second Life is used as the principal platform for analysis because it is one of the most popular 3DVEs in use globally. Since its creation in 2003, SL has stabilised in its usage patterns to a company-reported

794.000 repeat log-ins per month (Linden 2011b). These statistics are contentious due to self-reporting by the Linden Lab corporation (the creators of SL); given the highly corporatised nature of the platform, it is in Linden Lab's interests to spruik the popularity of the platform to remain appealing to investors and potential users. Linden Lab stopped reporting these types of statistics in 2011, probably due to falling numbers of users. However, Second Life's "Grid Survey" shows that of over 39 million total residents (many of these can be assumed to be unused accounts), there are commonly over 40,000 online at any given time (Shepherd 2014). Despite the uncertainty about the user-statistics, SL's corporatised environment closely replicates that of Western corporeal life, with just about every facet of life in SL commodified, as it is in the corporeal. For this reason, the body, gender, sexuality and space in SL are also discussed in relation to commodified experience, exchange value and desire. In addition to popularity, SL is a space premised on social interaction with a reliance on user-generated content. This highlights the synchronicity between hyper-capitalism, individualism and homogeneity; whilst possibilities are almost endless in terms of SL capabilities of spatial and avatar creation, there exists a surprising sameness with the corporeal even throughout the most diverse and subversive SL locales.

As a collaborative virtual environment (CVE), SL needs to be acknowledged as the virtual relative to a corporeal reality. In order to simplify some of the discussion, and to make it clear which environment is being spoken about, the two notions are at times separated in the analysis of SL spaces, so that corporeal existence is often referred to as CL (corporeal life) and the Second Life experience is referred to as VL (virtual life). VL is used here, rather than SL, because SL is used to refer to the platform itself, rather than the experience of the platform. In making this distinction, however, it is necessary to acknowledge that the corporeal and virtual are not entirely distinct in their creation, use or experience. The environments are different visually, navigationally and experientially, and so the terms "CL" and "VL" are used to denote this difference. However, there are probably more similarities than differences in how the two realms are experienced, due largely to VL being an extension of CL and including the same historical, cultural, economic and political influences as CL.

Second Life is less of a game and more of a real-time visually premised meeting space, where users interact, liaise and conduct business. It is possible to own virtual land and make virtual objects, including virtual money which can be converted to real US dollars. In many ways SL is a window into the future of CL, 'like a masquerade of "the real" [which] also functions as a laboratory for the development of digital, robotic, and artificial intelligence processes that may be commonplace later in our century' (Dethridge 2009, p. 263). As posited by Sermon and Gould (2011, p. 15), virtual environments are 'new forms of social narrative' whereby participants continually reconfigure their identity and ways of being in the world. Central to these individual narratives is the avatar, or personalised representation, which is normally humanoid in appearance. Avatars can range from the default selections given during the sign-up process of SL, to those that are highly customised and reminiscent of fantasy creatures. The avatar or avatars (it is possible for users to have more than one) are the principal means of expression in the SL environment. For users, the extent of investment into the platform can be measured in the personalisation of their avatar. As explained by Boellstorff, avatars are 'not just abstract anchors of virtual perspective; they [are] the modality through which residents experienced virtual selfhood' (2008, p. 129).

One of the greatest similarities between the corporeal and virtual worlds is shown in the importance placed on appearance and the visual aesthetic. Studies by Yee and Bailenson (2009) suggest an avatar's appearance is the primary indicator of how they are received by others, the way they interact, and in the case of games, their success as players. Such an emphasis on appearance in virtual environments means that many residents, in efforts reminiscent of CL, spend a good deal of time, money and effort in manipulating their virtual appearance to the point where 'embodiment was seen to reveal something deeply true about the choosing self' (Boellstorff 2008, p. 136). Research at Stanford University, headed by Bailenson, has revealed that the attractiveness of an avatar has implications for how an individual views both their online and offline selves (Dell 2008):

> [S]ubjects using good-looking avatars tended to display more confidence, friendliness and extroversion, just as in the real world: they approached avatar strangers within three feet, and in conversations tended to disclose more personal details. Ugly-duckling avatars, meanwhile, stayed five and a half feet away from strangers and were more tight-lipped.

Nick Yee's 2009 study found a direct correlation between gender and height, with male participants opting for taller characters and females opting for shorter ones. In addition, females were more inclined to choose avatars that were more attractive than their male counterparts (Yee, 2009). Similar results have been found by Vander Valk (2008, p. 207):

> for all of the attention given to the potential for positive – experimental, liberating, or emancipating – transformation of one's physical appearance and identity in virtual environments, it is just as likely that individuals will choose to focus on, and substantially reinforce, socially constructed notions of gender.

This phenomenon is experienced within SL as well as generally throughout the internet. According to Sermon and Gould (2011, p. 17), 'it would appear that the majority of Second Life users have chosen to accentuate the sexual signifiers of the perfect body'. It is the reducibility of the self to gendered expression, and therefore sexual agent, that is the focus of this study, which in so doing, will extend observations made regarding gender distinctions in the virtual realm.

The four SL regions that are referred to throughout this text were selected for analysis due to their relevance as regions:

- of supposed sexual diversity;
- that offer divergent tropes of identity and representation as related to notions of sexuality and gender;
- that are coherent with reference to the representations of ancient worlds, particularly those of Classical societies;
- that are popular, and therefore have a high number of visitors, given the previous conditions are met.[1]

The regions were chosen as spaces of potential gender and sexual fluidity that have their own rules, guidelines and norms. The four spaces are: The Lost Gardens of Apollo; Zeus; Eden; and Greek Gold. They were chosen because they have met the criteria of tolerating and/or celebrating sexual diversity as well as being tolerant of subversive identity representation, being representative of an ancient world, and being reasonably popular. The following overview provides a brief description of each space, including explicit rules, norms of governance and spatial characteristics.

Introduction to the four spaces

The Lost Gardens of Apollo is a visually rich island full of complexity and detail. Created by Dane Zander, the region is featured in SL's destination guide which in part explains its popularity. During the time spent researching Apollo, there were a number of instances where the space was too busy to visit, and so access was denied. Limitations on visitation numbers are due to the phenomenon of "rezzing," whereby the screen lags, making real-time interaction difficult. The region is rated G, general classification, which means it is forbidden 'to advertise or make available content or activity that is sexually explicit, violent, or depicts nudity' (Linden 2011a). The welcome note of the island (the pop-up that appears upon teleporting to a space) states 'No nudity, No RP (role play)' (Zander 2006) which reinforces the Linden classification and the non-sexual nature of the region. Although the welcome note states that the island is for 'all genders, shapes, creeds and colours' (Zander 2006), the avatars present throughout the research period were predominantly human in form, hyper-gendered, and attractive in terms of Western norms. There were also many avatars with the appearance of fantasy or mythological characters which emphasises the other-worldly and classical utopian feel of the region.

There are many places on the island that invite seclusion and it is very common for avatars to be seen alone and not interacting with others in many of the places on Apollo. The exceptions are Hyacinth Valley, where small groups of avatars can be witnessed practising Tai Chi, and at the Salsa y Boleros, where avatars slow dance together on a suspended platform overlooking a coastal section of Apollo. Both the Tai Chi and slow dancing can be performed by any avatar using the "pose balls" situated at Hyacinth Valley and Salsa y Boleros respectively. There are other pose balls located around Apollo in secluded places, on chairs or rugs for example, that animate avatars into cuddling each other. These animations are consistent with the sensual, but not overtly sexual, theme of the island.[2]

Zeus Gay Club and Concert Hall is space located within a region called Gay City Cologne, catering predominantly for gay males. It is part of a wider region entitled Gay City Estate which has entertainment as well as residential areas for the LGBTI community. There are no restrictions on who can visit the space, so female avatars can, and do, frequent it. According to its website, zeusgayclub.com, it 'is one of the most popular Second Life gay clubs, located on Gay City Estate owned by Michelangelo Villota' (Zeus Gay Club 2012). The space was founded by Franco Box, but he has since rescinded his ownership to Michelangelo Villota who is assisted in running the venue by a small number of club managers. According to the website, Franco Box's 'vision was to create a meeting point and a network for gay people in Second Life'. Unlike many other SL spaces, however, Zeus has no explicit rules or codes of conduct. On arrival at other spaces it is common to receive a welcome note, like that for Apollo, which includes guidelines for behaviour, including actions that are forbidden within the space. The absence of specific guidelines adds to the hedonistic feel of the space, similar to a holiday island experience in CL where there is an undercurrent of "anything goes."

Zeus can be categorised as an entertainment precinct, with its principal activities being dancing and music at the concert hall, shopping at the mall area, and relaxing in the surrounding areas that feature sports, social interactions and play typical of a seaside resort. It has a "moderate" rating, meaning that it is not a space explicitly created for sexual, or other adult-related, content. As a guide, the "M" rating is explained as follows (Linden 2011a):

> Residents in these spaces should expect to see a variety of themes and content. For instance, stores that sell a range of content that includes some "sexy" clothing or objects can generally reside in Moderate rather than Adult regions. Dance clubs that feature "burlesque" acts can also generally reside in Moderate regions so long as they don't promote sexual conduct, such as through pose balls (whether in "backrooms" or more visible spaces).

Although it is not an explicitly sexual space, there are many features of Zeus that categorise it as "mature". For instance, many of the shops contain items that are highly sexual, such as BDSM toys and clothing, and there are many billboard advertisements that feature sexual products, such as pornography and sexual services. As a large region, Zeus is a popular site in-world (with around 700 daily visitors according to its website), and features many theme nights, as well as live performances and language support for non-English visitors.

Founded by SL resident Yenzo Cortes, Eden is a space that holds nightly and weekly events, as well as having surrounds that include a dance club, a large shopping area, and a castle with secluded rooms. According to its welcome note and associated advertising, Eden is a space for 'lesbian and bi girls' and is part of a region called 'The Seduction' on SL. It is a popular space, hosting regular themed nights and events, and, according to its SL information page, has 'a community of 1,500 Residents and growing' (Greek Gold Lesbian Resort 2011). It is zoned as an "adult" area, meaning a region with 'expressly sexually themed content, spaces or activities, whether or not photo-realistic', where 'sexually themed' is defined as 'any sexually oriented activities and conduct' (Linden 2011a). Additionally, Eden has its own rules (Sabine666 Kuhn 2012):

1. Keep Eden a nice place
2. Keep peace, tolerate, and respect others
3. No weapons allowed to use
4. No extreme outfits (incl. chains, leaches etc.) Latex and collar is cool
5. No Nudity outside the designated areas
6. Please report violations to the next available Officer or Gatekeeper
7. Violations might result in a termination of membership [sic]

In addition to the general information found on the SL website, there is a 'Sisters of Eden' website and blog devoted to Eden and its activities.

The area which Eden takes up is relatively small, with all major attractions within a small distance of each other. It has an island-like feel and the appearance of a beach resort, including a beach with sand and umbrellas. The major activities on Eden, however, and the spaces where most avatars can be seen to congregate, are around its club, which often has theme nights and events, and its castle, which is used for sexual and intimate behaviour. Around the major structures of a castle and club are shops, a temple and various areas for lounging and viewing. The surrounds take the form of beaches, rocks and trees which add to the resort feel of the space, and also a mountain-like wall that separates Eden from the rest of the region, giving it a sense of privacy and exclusivity. As described in Chapter 4, the space has a tourist feel to it, in its resort structures and through its elicitation of the tourist gaze that evokes the discursive structures of sex tourism.

According to its founder, Greek Gold is the 'fastest growing lesbian SIM [region] on SL' with over 10,000 members (Greek Gold Lesbian Resort 2011). It is a large discrete island, far bigger than Eden, and is surrounded by a coast which gives the space an island resort feel. Its main features include an entry space where avatars are greeted by a Greek Gold administrator, a large dance club, coastal lookouts, gazebos for private retreats, a number of waterfalls overlooking a river in the central part of the island, a margarita bar and a beach for surfing. The space is inclusive of all women, explicitly stating 'lesbian, dyke, butch, boi, femme, girl, (and) gay' females (Greek Gold Lesbian Resort 2011). Its creator Hera Emms, is, according to her webpage, 'mistress to six beautiful girls' and 'practices BDSM and the Greek life' (2011). The welcome note received by all on arrival to the island lists specific rules for conduct (Hera Emms 2011):

- No bio males – no exceptions to this rule
- No poofters or particles on the dance floor
- No nudity (all privates must be covered, please note, we respectfully request that butches/bois keep their nipples covered – pasties are permitted).

The 'No poofters or particles on the dance floor' rule refers to the prohibition of animated objects being activated in the space of the dance floor. This is due to the limited capacity of the platform to handle large numbers of moving objects in a single space.

As is the case on Eden, Greek Gold prohibits males from visiting the island. Importantly, it is framed as 'bio males', which suggests that transgender individuals are welcome, along with varying depictions of masculinity as represented by the base gender of female. In addition, there is a 'no nudity' rule, which positions the space as a non-explicitly sexual environment. To offset the potential confusion of masculine females showing their nipples, the guidelines are that 'pasties' (usually appearing as a small plaster place over the nipples) can be used. This rule, in particular, reveals the complexities of representation of the masculine female form: although many butches and bois have the appearance of a flat-chested male, they are only permitted to go topless if wearing pasties. This underscores the emphasis placed on female modesty in Western culture, whereby the female body is essentially sexualised, even in the absence of curves, and the breasts actively displayed as feminised sexual objects.

Chapter overview

Chapter 1 provides a theoretical overview of the key literature that informs this work, introducing the concepts of embodiment, virtuality, cyberspace and subversion. Collaborative virtual environments are framed as cultural phenomena that can be conceptualised principally as social systems of interaction, communication and signification. In an environment centred on the visual, embodiment is a crucial phenomenon for engagement and is seen as the key form of signification and social access. Social interaction and representation are based on the corporeal, and it is in the everyday physical world that such signifying processes are founded. Both words are inextricably linked, and so the practices of identity formation, selfhood and expression are spoken through the discursive power structures of the corporeal which inform notions of normative, subversive and marginal identities and practices. This overview of signifying practices and virtual embodiment is framed by queer and feminist theories, looking to identify and expand upon the social conditions of life both in front of and through the screen. What at first seems an unlimited and unrestricted platform for creation and imagination is, after scrutiny, revealed as the entrenchment of our social and cultural conditions.

After establishing the major theories about the body and space, in Chapter 2 the discussion progresses to the specific discourse of tourism. The SL regions studied are framed as tourist spaces via the dual motivations of sunlust and wanderlust, which in turn lead to the conceptualisation of experience as either centred on bodily pleasures (liminoid) or focused on meaning and engagement (liminal). Seen through the lenses of engagement and meaning, the notions of authenticity and inauthenticity of experience in the virtual environment are broached, particularly in relation to the historical narratives offered in each of the SL spaces. Emmanuel Kant's concept of the sublime is also evoked here in reference to the natural, the technological and the digital sublime, and it is argued that all three can be attributed to certain regions. Given the focus on the spaces as sites of pleasure, in addition to a more general emphasis on the body within SL, the gaze is analysed in terms of the pleasures and politics of looking. Furthermore, with a focus on queer identities, the spaces are read in terms of the pleasure periphery as it relates to the coastal holiday islands and paradise narratives of the Mediterranean.

Chapter 3 continues the discussion regarding space, the historical world and the Mediterranean environment but extends it to incorporate a more specific analysis of gender and sexuality as related to the particular space of The Lost Gardens of Apollo. As the most visually spectacular of the regions studied, Apollo illustrates the concept of the sublime as applied to 3D virtual environments as well as illuminating the paradigms of male beauty and utopian space specifically as read through the mythology of ancient Greece. Through this lens, Apollo is discussed as a masculine environment and raises questions regarding the reading of sexuality as represented by and interpreted through the male subject. The hyper-masculinity displayed by the avatars throughout the space is discussed as an idealised form of male representation, which subsumes the homosocial and homosexual within it. Although subversion can be read into the space of Apollo, through its historical narrative and explicit references to non-normative sexualities and representations, normativity is reaffirmed through the prevailing standards and conduct required of its visitors. Heterosexual coupling is normalised by the displays of affection permitted and the hyper-gendered representations of avatars.

The gay male space of Zeus is analysed in Chapter 4 and extends the discussion of masculinity and sexuality. Whilst Apollo is a region of spectacular natural landscapes, Zeus by contrast offers a space typical of contemporary island tourism. It is a space premised on pleasure and the body as a consumable product. What is on display here, and what can be appropriated in terms of interaction, is in stark contrast to the new conservatism of contemporary Western gay culture. On Zeus, the hyper-masculine male form is both normalised and commodified in terms of its sexual exchange value. Signified through markers of masculinity, the male subject raises questions regarding the moral panics surrounding subversive male sexualities and their marginalisation to spaces of deviance. Heteronormative and homonormative representations are analysed in reference to consumerism as well as the movement towards conservatism of gay male identities in Western society, which allows for the legitimisation of certain subject positions over their alternatives.

In Chapter 5, the discussion turns to female representation and sexuality. The positioning of the body as commodity, seen in the case of the male body in Apollo and particularly Zeus, continues in the space of Eden which exemplifies the positioning of the female body as consumable. Whilst the male body and sexuality in its hyper-gendered state is often read as threatening, the sexualised and hyper-gendered female body often signifies temptation and vice. Although positioned as a female-only space suitable for queer-identifying women, Eden is highly heteronormative. The male gaze is inscribed into the viewing practices of the space, leading to the reading of Eden as potentially voyeuristic and pornographic in reference to hetero-patriarchal norms. Femininity as attached to the female subject position is normalised alongside the body as commodity, its usefulness attributed to sexual exchange value. The spatial features of Eden are also analysed in terms of how they contribute to hetero-patriarchal usage and practices. Furthermore, Eden is discussed in reference to its position as an island paradise as well as by references to Biblical and symbolic significations.

In contrast to the heteronormative representations on Eden, Greek Gold offers a more diverse spectrum of gender identities. Chapter 6 presents the region as a study of female masculinity, in many ways in direct contrast to the feminine and sexualised displays of Eden, This is not to say that representations on Greek Gold are not sexualised; however, the diversity that is displayed disrupts much of the signification as attributed to the female body. The chapter relies heavily on Judith Halberstam's work on female masculinity as it analyses digital representations of the female body outside of female femininity. The plurality of expression in SL, branching across the masculine spectrum, detaches masculine attributes from the male body even more so than in corporeal signification practices. The destabilisation is at the site of the desired self and reveals that the female body can be desired as masculine as well as feminine, in both object and subject positions. As is the case with Zeus, however, Greek Gold is ultimately a subversive site that operates against the normative.

1

Embodiment, Virtual Experience and the Body: Possibilities for Subversion?

Introduction

The focus of this chapter is to contextualise notions of the body as well as the space it inhabits. The body, seen from the discipline/s of cultural studies, is a political object that has inscribed onto it history, culture, society, sexuality, violence and power. It is our central point of understanding of both ourselves and the outside world, and so carries with it both history and the present, as well as the future. It is also through this field of enquiry that both feminism and queer theory are situated vis-à-vis each other, often in uneasy acknowledgement of the politics and usefulness of the other (whilst the fields of theory often disagree about the scope of fluidity of the gendered and sexed self). Whilst cyberspace and virtual worlds may seem initially to be very different forms of space from those experienced in the corporeal, they operate in much the same way, using the same foundational codes and conventions of the real. This is an extension of Lefebvre's, and other spatial theorists', contentions that space is produced through ideological investments, and thus reinforces discursive power structures. In addition to embodiment, gender and sexuality are considered as having spatial dimensions, which produces tensions within a world built on fantasy that looks to offer alternative notions of equality and engagement. The SL platform itself, being user-generated, appears as a promising platform for alternative spatial construction, and therefore spatial discourse, by permitting different forms of embodiment as well as having no established rules for building, constructing and articulating space. However, and as noted by Joseph Clark (2011, p. 145), this potential is largely unfulfilled:

> Even though the resulting spaces often represent entirely fantastic or fictional worlds, cities, and landscapes, at their most fanciful

> they still tend to contain at least a modicum of the world as we know it. Buildings, highways, parks, seacoasts, and forests in these virtual worlds may be utopian or not physically possible in real life, but they are still, like any other creative endeavour, reflective of the hopes and dreams – and the bias and ignorance – of their designers. They perform ideologies just as sculptures, movies, monuments, and buildings do in real life, and are thus rhetorical constructions.

Corporeal life proves to be inescapable in imagining alternative realities. Apollo, Zeus, Eden and Greek Gold offer less transgression than what is initially imagined in visiting subversive spaces. Although some of the features of the four regions defy physical constraints, such as occasional buildings, sculptures and monuments, the layout, architecture and design of each region largely mimics those either in existence in CL, or that from an imagined past. In re-presenting the ancient world, the four regions signify a cybertopia that exists beyond the established parameters of contemporary existence, yet simultaneously evokes the collective imagination of a utopian past. However, this imagined space is inextricably linked to the corporeal and so is unable to depart from it in any significant political or cultural sense. What is meaningful, however, is the engagement with space for the individual, whereby some form of authenticity can be achieved (this notion of authenticity, particularly applied to tourism, is taken as a subjective phenomenon that is based on the experience of the individual, rather than any objective claim to a truth or reality as authentic).

The body (embodiment)

The term 'embodiment' can have varied and sometimes conflicting meanings depending on the field of enquiry and the background that is drawn upon. It is often a term that is taken for granted; we all have a body, therefore our embodiment is implicit. However, when exploring the ideas of alternative embodiment, and bodies that extend beyond those of the traditional/normative, it is a term that necessitates an explicit definition. For the purposes of this analysis, the more useful definitions are drawn from the social sciences. Ziemke identifies the notion of 'social embodiment', initially developed by Barsalou, Niedenthal, Barbey and Ruppert, which 'addresses the role of embodiment in social interactions rather than the question of what type of body is required for what type of cognition' (Ziemke 2003, p. 1305). This account of embodiment differs from others in its focus on social interaction rather than biological processes. That embodiment is commonly defined within the boundaries of physical existence is not surprising; it has only been recently that the body, or a body or bodies, in relation to the mind or self, has been re-examined in light of anthropological, psychological and technological advances. This shift in scholarly thinking suggests that 'the body is passing through a critical historical moment ... [that] offers a critical methodological opportunity to reformulate theories of culture, self, and experience, with the body at the centre of analysis' (Csordas 1994, p. 4). Principally, there has been a rejection of brain-bound notions of identity and self-formation and a shift to analysing the brain, the body and the world, wherein the body's interactions within the world, and perceived by the mind, are the essential features of embodiment and cognition.

Poststructuralism views the body, or our conceptions of the body, as having a history. According to Csordas (1994, p. 4), the body 'is as much a cultural phenomenon as it is a biological entity', and the value of signification is central to defining bodies. He suggests that 'culture is grounded in the human body' and this is ultimately the site of meaning in terms of our selves and how we make sense of each other and our environment. Mind, body and environment constantly interact, being inseparable from social interaction and culture, so that 'the full explanation of our knowledge of self arises from the participatory interaction with our embodied existence' (Gibbs 2006, p. 17). We gain meaning, and a sense of self, not from looking inward, but from looking outward in projected reflection of others, what Sartre describes as '*being-for-others*' (1943, p. 305), suggesting that 'I must apprehend the Other first as the one for whom I exist as object' (p. 339). If the idea of body and self are dependent on reflection of and interaction with others, then they are constructions of interaction and thus constructions of culture (Gibbs 2006, p. 13):

> Bodies are not culture-free objects, because all aspects of embodied experience are shaped by cultural processes. Theories of human conceptual systems should be inherently cultural in that the cognition that occurs when the body meets the world is inextricably culturally based.

Embodiment is therefore subject to cultural influence, categorisation and construction, as the body-in-world is shaped socially and ultimately politically. Vygotsky articulates this premise by contending that social processes 'operate according to sociological and economic principles, particularly the principles of exchange value and commodification' (cited in Wertsch 1985, p. 60). These principles of exchange and commodification have a profound influence in consideration of body-as-object and the body as performative.

The way the body (a body) is lived and viewed is central to experience, and embodied experience has crucial effects on the individual in terms of identity and the formation of self. Often in the consumerist, capitalist society in which we live, the body is experienced as an object; it is displayed and performed, consumed by others. Consequently, this objectification has an experiential effect on the mind and the perception of the self, as one's experiences of the body-on-display impacts significantly on one's sense of worth in society. Csordas suggests that the modern world's emphasis on the body as object has had the effect of 'the individuation of the psychological self and the instantiation of dualism in the conceptualisation of the human being', brought about by 'reflective, ideological knowledge' (1994, p. 7). This objectification contrasts with viewing the body as subject, where the mind and self are aware of a bodily occupancy in the world. The political body is centred around body as performance: that which has a history of performing and being performed upon. The performative body is indelibly linked to the enculturated historical body, which is ultimately linked to the political and struggles of power. The historical body, as is true for any cultural artefact, contrasts 'the important and the subordinate, the essential and the accidental, plans and accomplishments, preparations and declines. These vectors which are traced through the dense whole of facts have already distorted the original reality in which everything is equally real' (Merleau-Ponty 1964, pp. 193–194). Body-as-object is the body as constructed through the contestations of history and cultural exchange, often violently forged. Such violence is particularly apparent in bodies of difference, what Leder refers to as 'the threatening body', where 'the body, forgotten in its seamless functioning, comes to thematic attention particularly at times of breakdown or problematic operation' (1990, p. 127). Dysfunction and difference in this description characterise a body at odds with the norm prescribed by the dominant political, social, economic and historic order, and include 'divergent' forms of gender and sexuality.

The cultural regulation of body-as-object produces normative performance. Corporeal displays outside of this normative order are labelled immoral or divergent: 'a problematic body' (Leder 1990, p. 135). Thus, controls placed on the gendered and sexed body are integral to embodiment, as both body-as-object and body-in-world. The body that interacts with, and is viewed by, others is imperative to an individual's consciousness and concept of self, so that bodily performance outside the norm has profound effects on embodied experience of both the 'divergent' body and the bodies of those who it interacts with. This is because the self understands itself through the reflection of others, those whom the self interacts with: 'the Other accomplishes for us a function of which we are incapable and which nevertheless is incumbent on us: to see ourselves as we are' (Sartre 1943, p. 354).

The current period of late capitalism and commoditisation has seen a new emphasis placed on the body, and while the era has seen a greater importance placed on individualisation, commercialisation often sees the paradoxical effect of homogeneity and the implicit regulation on appearance and bodily performance (Csordas 1994, p. 2):

> In the milieu of 'late capitalism' and 'consumer culture', with its multiplicity of images that stimulate needs and desires and the corresponding changes in material arrangements of social space, the body/self has become primarily a performing self of appearance, display and impression management.

Where, historically, individuals were controlled explicitly through the methods of the state apparatuses of the health, school, political and judicial systems, in late capitalism individuals regulate themselves through the implicit standardisation of norms, where the body has 'acquired new importance as tokens of class and status distinctions' (Turner 1994, p. 28). Alongside the pressure to maintain a performative body marked by the new class and status order, the body is strictly regulated in performance of gender and sexuality where strict codes remain for the expression of acceptable embodiment, a phenomenon explained by Butler (1993, p. 12):

> Performativity is thus not a singular 'act', for it is always a reiteration of a norm or set of norms, and to the extent that it requires an act-like-status in the present, it conceals or dissimulates the conventions of which it is a repetition.

Embodiment, therefore, in both virtual and corporeal spaces is a cultural expression of societal norms placed on the body and regulated through accepted modes of interaction and display. In the virtual environment, this interaction occurs through modes of representation, so that identity is performed through the body, as it does in CL.

Capitalist objectification has ramifications on bodily performance, codifying displays of body-as-object and body-in-world regulating gender, sexuality, race, ethnicity and various forms of bodily 'error' such as disability, deformity and illness. Much work in feminist and queer theory has focused on the forms and apparatuses of the regulation of gender and sexuality, whilst others seek to explain demarcation as historical and cultural constructions that exist only in subjective interaction. Butler asks (2004, p. 41):

> Is there a gender that pre-exists its regulation, or is it the case that, in being subject to regulation, the gendered subject emerges, produced in and through that particular form of subjection? Is subjection not the process by which regulations produce gender?

Butler questions the history and supposedly natural properties of sex and gender, undermining biological determinism and the objective truth of scientific discourse. This closely aligns with the work in psychology regarding the cultural grounding of embodiment: that of the body-as-object and the performative aspects of the body as central to cognition, consciousness and the formation of self. The 'self' is as much a product of culture as the body since there is no self that is not immersed in some form of experience or situation: 'To be human, indeed to be living, is always to be in a situation, a context, a world' (Varela, Thompson & Rosch 1991, p. 59). Embodiment, as the principal condition of interaction with the world, is key to defining the self and identity. It is the feature that reveals the self to the world and is central to self-reflection, a constant arena for self-improvement, and is the feature that creates the most meaning in cultural exchange.

Embodiment and the self reflected through lived experience produce affective states that have consequences for behaviour and interaction. Moore and Isen (1990, p. 18) contend that 'the effects of feeling states on self are mediated through perception and memory', which 'affects reactions to self and others' where perception is intrinsically linked to an individual's embodied state. Gibbs explains that 'rather than being a biological given, embodiment is a category of sociocultural analysis, often revealing complex dimensions of the interactions between bodies and personhood' (2006, p. 37). Experience, grounded in culture, has profound effects on both emotion and body modality where both act and react reciprocally. Embodiment is a dynamic process: it is dynamic because it is subject to internal and external reflection, and re-construction, and it is a process because it involves continuous boundary negotiations. The body can extend past itself, either by incorporating tools, or by utilising the virtual to have several bodies, realising the multiplicity of identity. The possibilities of virtual embodiment may be infinite, but what is certain is that *a* body, in whatever form it may take, is an essential prerequisite. While we may extend the boundaries of corporeal possibility, such bodily potentials must have a reference point, a point grounded in history and culture, to work from. For this reason embodiment, even in the virtual realm, is subject to the same history, culture and politics of the corporeal.

A key element of feeling both connected to an environment and others is feeling connected to one's own body. In virtual worlds, the extent to which a person feels present, attentive and capable of action in an environment is referred to as presence. It is a concept that is closely linked to immersion, 'a sense that sensory experience of the actual world [is] sufficiently muted, and sensory experience of a virtual world sufficiently heightened' to the point where an individual no longer feels a part of the corporeal world (Boellstorff 2008, p. 112). By heightening the active senses and dampening the others to the degree that they diminish into the background, virtual programs can increase the level of presence that the user feels. Biocca describes the sense of presence as oscillating, whereby the user's experience 'can be interrupted by sensory cues from the physical environment and imperfections in the interface' (2006, p. 9). However, as noted by Boellstorff, immersion is not necessarily the goal for three-dimensional virtual environments (3DVEs), nor is it essential for interaction or participation. The world of SL is predominantly social, and so the environment can take on a similar scope to those of other social media, whereby a user may flit between active and passive online states as they move between their corporeal and virtual existences. Denise Doyle, in discussing avatar bodies in SL, refers to this experience as a mixture of presence and absence, or 'a mix of objective "looking" and a subjective sense of "being"' (Doyle 2011, p. 105). However, whilst individuals may use some SL spaces purely for social interaction, there are certainly regions that are visited for their spatial features and ability to facilitate escapism, as is the case with the four regions studied. Within environments that rely on the visible as the key function for presence, it is the spatial aspects of the SL spaces, including layout and features, in addition to the ability to move one's avatar – effectively being in control of one's embodied state – that affect how immersed and present an individual feels.

Feminism and queer theory: Positioning the body in cyberspace

Cyberspace, of which CVEs are a part, can be defined as the social networks of interactions rather than just the technical means of the network itself. This definition leads to conceptualising cyberspace as a cultural medium, which is something that is primarily visual in how it is experienced. Refuting the notion of technological determinism, culture is posited as the driving force behind the various uses and appropriations of the internet specifically, and technology generally. In discussing the culture of cyberspace, Robins (1996, p. 137) asserts that the 'institutions developing and promoting the new technologies exist solidly in this world' and so we cannot discuss technology and virtual worlds as if they have no reference to the lived world, with its politics, culture and structures of power. It is not possible to fully analyse virtual worlds as if they were somehow detached from our day-to-day existence. Whilst the possibilities may be endless in virtual worlds, we must view the technology 'not as outside society, as technological determinism would have it, but as inextricably part of society' (Wajcman 1994, p. 6). Technology and cyberspace are bounded by the politics, conventions and male-dominated ideology from which they were formed.

Gender has long been thought of as something natural and distinctive. Through science and philosophy, areas of study dominated by white, heterosexual men, this position has been reinforced: 'an essentialism that assumes that there is universal and underlying order that can be explained through "biology" and "natural" characteristics' (Grosz 1995, p. 47). Recent work by feminist theorists in science and philosophy has overturned preconceived ideas, and with varying degrees of certainty, shown that much, if not all, of what is thought of as a natural in terms of masculinity and femininity is constructed by society and culture. Physical codes of appearance, movement and behaviour are central to the construction and perception of gender and sexuality in society. The body 'is a powerful symbolic form, a surface on which the central rules, hierarchies, and even metaphysical commitments of a culture are inscribed and thus reinforced through the concrete language of the body' (Bordo 1997, p. 90). The notion that gender codification is culturally constructed is elaborated upon by Judith Butler who explains, 'even if the sexes appear to be unproblematically binary in their morphology and constitution ... there is no reason to assume that genders ought also to remain as two' (Butler 1999, p. 8). In addition, Butler also views the basis of sex division and differentiation as a construction, and 'as the effect of the apparatus of cultural construction designated by gender' (1999, p. 11).

Cyberculture and cyberbodies can facilitate a new form of embodiment and new ways of inscribing and interpreting the self that are central to the reconfiguration of postmodern identities. Not only is gender reconfigured as a construction of the Enlightenment, but also sexuality in the form of compulsory heterosexuality and the assumption of 'naturalised' relations between men and women. From this perspective, gender and sexuality in their traditional conceptions can be viewed as performance, where 'gender inscribes the fantasy of a true, stable gender identity on the surface of the body, which simultaneously naturalises the normative fiction of heterosexual coherence' (Sunden 2003, p. 53). However, the dominant culture has effects on identity construction and maintenance in cyberspace. Robins suggests that identities 'are composable insofar as the constraints of the real world and real-world body are overcome in the artificial domain' (1996, p. 138), which may be no easy task considering how many ingrained social and cultural constraints are inscribed on the body and its uses. As previously mentioned, physical and sexual difference have historically been conceptualised as something natural and biological, so to overcome these constructed differences in the virtual world would require an individual to distance their online existence markedly from their real-world reality. For example, in the case of gender it is often assumed that an online user is either male or female. Options for joining various games, social networks and virtual worlds have it as a requirement that one sign on as a predetermined gender. This has the effect of compartmentalising male and female roles before one even enters the world, undermining the ability to fluidly embody an ungendered, transgendered or gender-ambiguous identity.

Furthermore, the culture in which technology is conceptualised is predominantly masculine. In the case of cyberculture, there is a history of male-centred design and use of hardware and software systems (Dholakia 2007, p. 237). This has had the effect of affirming men's 'masculinity through technical competence and posit(ing) women, by contrast, as technologically ignorant and incompetent' (Wajcman 1994, p. 11). It is only recently, with the advent and popularity of social networking sites, that women have found a more equal usership for internet usage (Dholakia 2007, p. 232). Much of the internet in its contemporary form is tailored towards the capitalist marketplace, where hardware and software are made specifically for niche markets that can be exploited for profit. In embracing female usership on the internet, companies are focused on revenue rather than inclusionary social or political practices. Those outside the key demographic of techno-authorship and usership are included in cyberspace only on the basis that they participate within the patriarchal capitalist paradigm. It has been suggested that utopian visions are mostly promoted by those who seek to gain the most capital from extra usage, presenting cyberculture 'as a source of salvation from the reality of a lonely culture and radical social disconnection from everyday life' (Kroker 1996, p. 168). It is important to remember that cyberculture is as much a capitalist enterprise as anything in the 'real' world, and that similar, if not identical, social, political and economic forces are engaged, regardless of the creative capabilities and possibilities of the virtual.

In addition to a feminist critique of the SL spaces, queer theory is utilised as a complementary discourse for analysis. These two perspectives are interwoven through the various chapters of this text, in the acknowledgement that 'such categories or forms of living are interrelated and at specific points for particular political and social reasons one may be more important in framing life and demanding political action than the other' (McLaughlin, Casey & Richardson 2006, p. 4). There are a number of reasons for using both theories simultaneously, and as suggested by McLaughlin, Casey and Richardson, at times one theory is better suited to the political outcomes and framing practices of various social and cultural phenomena. The below categories are illustrative of this:

Sex	Gendered and commodified (feminist) Subversive and non-normative (queer)
Identity	Masculine/feminine binary (feminist) Trans, androgynous and non-normatively gendered (queer)
Power and Agency	Capitalism, domesticity, the controlled body (feminist) Body as expressive, subversive (queer)
Cyberspace	Technology as masculine (feminist) Technology as emancipatory – virtual embodiment (queer)

Furthermore, Diane Richardson states that 'A primary focus for feminist writers has been on how (hetero)sexuality is related to the maintenance of male domination and gender hierarchies, whereas within queer theory attention has been on the ways in which "heteronormativity" functions to privilege and sustain heterosexuality and exclude sexual "others"' (Richardson 2006, p. 37). This book aims to throw some light on both ideas, so that heterosexuality and gender are critiqued in view of male domination, female sexuality and capitalism, and that heteronormativity and the queer subject are discussed through the tropes of deviance, subversion and the new conservatism of queer.

Of particular importance is the tension at times ratified between feminist and queer discourses that sees the gender-identified individual as antagonistic to the goals of feminism in reference to the dissolution of power difference and sexual practice. For instance, Judith Halberstam discusses the female androgynous and transgendered individual as gaining agency through the masculine paradigm, which may be considered as undercutting the feminist goal of equality through the re-enactment of masculinised behaviours and identities. Whilst acknowledging the critique of some feminists that such gender identification reinforces gender binaries, I believe this is too simplistic a reading of the androgynous and trans-gendered subject because of the little affordance this critique allows for gender diversity and fluidity. What the critique does not allow for is the definitive separation of gender from biological sex, and the ensuing performative tropes of gender itself that cross biological sex categories, ultimately destabilising the power structures of sex and gender, that is not often recognised in sole feminist analysis. So it is argued that rather than reinforcing the power structures embedded in binary gender, the androgynous, trans and fluid subject can in fact destabilise these categories.

It is these subjectivities which question the naturalised order that are labelled as subversive. As Butler acknowledges, what constitutes the subversive is not a fixed category of being or knowing. Like all constructs, it is in a constant state of redefinition with that which is defined as normal and 'unsubversive'. Moreover, the context in which the subversive operates needs to be acknowledged as an essential aspect of meaning (Butler 1999, p. xxi):

> Just as metaphors lose their metaphoricity as they congeal through time into concepts, so subversive performances always run the risk of becoming deadening clichés through their repetition and, most importantly, through their repetition in commodity culture where 'subversion' carries market value.

What the 'subversive' does do, however, is function as a counter-point to some reality, or some naturalised quality that is called into question (Butler 1999, p. xxiii):

> And this is the occasion in which we come to understand that what we take to be 'real', what we invoke as the naturalised knowledge of gender is, in fact, a changeable and revisable reality. Call it subversive or call it something else. Although this insight does not in itself constitute a political revolution, no political revolution is possible without a radical shift in one's notion of the possible and the real.

The line between cliché and the questioning that subversion ought to do, as defined in this work, is at the level of insight generated through the questioning of assumptions about the naturalised conditions of gender and sexuality. The heteronormative portrayals of female-to-female sexuality on the site of Eden are read as clichéd performances that reiterate normative positions. However, the island of Greek Gold is read as subversive because of the representation of fluid gender that does not subscribe to normative assumptions, and in fact calls these assumptions into question.

One of the key aims of this book is to use the subversive as applied to gender and sexuality (through both feminist and queer theories) to destabilise traditional power structures of patriarchy as well as heterosexuality and heteronormativity. Moreover, queer theory works with the notions of virtuality as well as the concepts of cyberspace, cyber-identities and cyber-embodiment to further destabilise the biological determinism of sex and gender. Extending Haraway's feminist reading of science, technology and the cyborg, which provokes a re-reading of gendered identity and power, queer theory can simultaneously re-work notions of sexuality, subversion and difference. In deconstructing the barrier between human and machine, Haraway fractures the boundary between what has been historically, scientifically and politically held as natural, and the reality of a hybrid, multi-faceted form of existence and individuality. The cyborg is central to the dissolution of boundaries, as it deconstructs the social, historical and scientific determinism that has dominated discourse since the Enlightenment era, and can be central to women's experience and liberation politics as the boundary between science fiction and reality becomes blurred. Similarly, the cyborg can be seen as outside normative sex and sexuality, being at once outside of sexuality yet indeterminably infused with human corporeal desire. In taking the physical body out of sexual embodiment, the subject is pre-designated as queer insofar as the normative conditions of sex have been subverted. There are no naturalised conditions of sex for cyber-bodies, no essential reproductive capabilities or responsibilities; cyber-embodiment is shifting, temporal and liquid and so are its practices. Sexuality, and sex itself, is removed from the essentialised referents of the corporeal, meaning that what is projected through the screen is a notion of culturally shaped representation.

Space as gendered

Both the production and conception of space have been historically, economically and politically the domain of men. In considering spatial production, and the usage of the terminology space and spatial, it is necessary to consider the epistemological framing of space as cultural and, therefore, patriarchal. Prior to the work of theorist Henri Lefebvre and other Marxist geographers of the twentieth century, space was considered more or less an objective entity, an apolitical and ahistorical element of the universe. In drawing upon Marxism, Lefebvre and his contemporaries challenged this idea, and instead postulated that space was in fact subjective with far-reaching social, cultural, historical and political ramifications. Marxist geography 'revolved around the need to recapture social control over the production of space from an expansive capitalism and an equally expansive and instrumentalist capitalist state' (Soja 1989, p. 46). And, according to Lefebvre, 'The very survival of capitalism ... was built upon the creation of an increasingly embracing, instrumental, and socially mystified spatiality, hidden from critical view under thick veils of illusion and ideology' (as cited in Soja 1989, p. 50). In view of this new critique, space was found to have as many social layers as human interaction; space was indeed built upon human input and relations, and driven by social processes. Soja (1989, p. 78) describes the socio-spatial dialectic as inter-related and in constant flux:

> The structure of organised space is not a separate structure with its own autonomous laws of construction and transformation, nor is it simply an expression of the class structure emerging from social (and thus aspatial?) relations of production. It represents, instead, a dialectically defined component of the general relations of production, relations which are simultaneously social and spatial.

Space here is conceptualised as a patriarchal construct, as part of the capitalist and modernist paradigm, whereby it is produced according to relative political and cultural discourses. Space, therefore, is both produced by and for patriarchal interests that serve the hegemony, as illustrated in the core and the periphery in usage patterns such as the work/domestic binary. As an extension of Marxist geography, feminist geography seeks to position women in wider social and cultural contexts that encompass class conditions, but also extends this to relations outside of the privileged work (economic)/domestic (social) binary. This extension is particularly important in conceptualising the lives of queer (and other) identities that do not necessarily fulfil the ideological functions of gender described by the white, bourgeois Marxist model. In deliberation of these issues, much of the analysis in this text is relevant to space more readily applicable to tourist and leisure discourse than that of work and/or domesticity. Thus, lived experience needs to consider space in a wider paradigm than those of economic relations, although analysis of these relations is a necessary component of agency within SL, as well as of tourist discourse itself.

As with space, technology is primarily a masculine domain. In writing of the technological sublime, David Nye remarks, 'At first the sublime was understood to be an emotion exclusive to the male sex, women being relegated to the realm of the beautiful, as in Kant's earlier writings' (1994, p. 283). Such exclusion has reduced the potential for a feminine technological discourse and spatial configuration of the world. Nye remarks that feminisation of technology occurs upon widespread adoption, consumption and domestication; however, such feminisation is *post factum* and, in effect, leads to the entrenchment of masculinised technological discursive practices. For cyberspace and the virtual, this has meant that both the production and conception of space as technological artifices are continuations of the corporeal discourse of the masculine. Although there is scope within SL to produce worlds outside of the prescribed spatial conventions of capitalist conventions, the spaces analysed here all reflect an adherence to the dominant spatial discourses regarding architecture, movement strategies, the 'natural' environment and the layout of zones such as shopping, private and entertainment locations. The virtual spaces analysed in SL are lived spaces that cannot be removed from corporeal political, cultural and spatial context. The production of virtual space in the four regions, despite the possibilities for re-imagining space, is consistent with the physical world. For all its possibilities, technology is bounded by a lack of imagination – there is little want for imagination in a realm that simply maintains its pre-existing power structures. A radical shift in technology and space would require a radical shift in the organising practices of corporeal life.

Inscribed into the four regions of Apollo, Zeus, Eden and Greek Gold are the foundations of historical, cultural and political technological and spatial practice. Whilst it is unsurprising that the male-dominated regions of Apollo and Zeus reflect a masculinised spatiality, it is perhaps surprising that the female-only regions of Eden and Greek Gold also reflect these practices. All four spaces contain areas of: landscapes of the sublime; structures of the sublime; beaches; secluded areas for intimacy; dance-floors; and shopping precincts. Such features run parallel with CL consumer and leisure spaces and practice, and no region makes a significant departure from them. As discussed in Chapter 2 (tourism), this spatial structure reinforces practices of looking, practices of movement and discourses of experience that place the body as central to expression, interaction and evaluation. The self in these four SL regions is not reified as having disembodied cyborg potential, but reinscribed as a desired virtual extension of the corporeal. The spatial configuration ensures an attachment to the corporeal by limiting the imagined potential for radical experience outside hetero-patriarchal capitalist existence.

Furthermore, this normalisation suggests a fear of an unknown future, or 'a crisis of unified white masculine subjectivity' (Fuchs 1995, p. 282) in reference to the disembodied subject. As a reaction to perceived crises of gender, and threats to the heterosexual matrix that seem ever more possible as those in the first and developing world progress through history and technology, there is almost always a negating force seeking to overturn potential radicalism of existence and future possibilities. The conservatism and hyper-capitalism which have spread throughout queer culture are symptomatic of a once radical subjectivity being subsumed by the greater culture through a methodical process of normalisation. That subjects in Apollo, Zeus and Eden (for the most part) embody hyper-gendered representations further exemplifies this rejection of a potential cyborg, or un-gendered, subject position as well as of a fluidly gendered and/or androgynous position. Whilst Greek Gold is representative of a space of difference in regards to gendered identity, showing a range of female embodiment from the feminine to the masculine, it must be noted that these often conform to the conventional structures of space and capitalism. What is written into the spaces, in repudiation of cyborg un-gendered subjectivity, is the notion of desire that cannot be imagined in the absence of corporeal being, so that the hyper-gendered identity in Apollo, Zeus and Eden operates through a highly gendered and corporeally likened sexuality. Even in Greek Gold, which initially seems to flout normative desire, there is some replication of conformity in fluidly gendered positions precisely because they are anchored to a female subject position (that this attachment is not, and perhaps cannot be, overturned is due to the pervasiveness of gendered identity within our culture).

Experiencing femaleness in Greek Gold is both liberating and confining, illustrating the complexities and paradoxes of space itself. As noted by the feminist geographer Gillian Rose, 'There is a notion of things that are not representable in masculinist discourse, but which women themselves may sense if not articulate' (1993, p. 138) insofar as there is a need for spaces that function outside of those already prescribed. Attempting to articulate such space in practice, however, illustrates the difficulty in operating outside of the given discourse. The spatiality of Greek Gold is a tacit replication of commonly imagined leisure space. It is dominated by its large dance-floor where most avatars can be found, but also includes various other leisure sub-spaces such as beaches, bars and secluded areas. It has been crafted for the tourist gaze as much as any heteronormative environment, yet it exists outside of it. There is an importance placed on the body, yet it is not for an assumed male gaze that operates omnipotently throughout space in general. There is also the notion that the space exists for a specific subculture in a small region in SL in cyberspace, so that its very marginality precludes its importance as a space at all. This is what Rose refers to as confinement, in that women's lives become pressed into irrelevance. The inability of women to display the freedom of gender fluidity and control of space exercised on Greek Gold in the corporeal reflects the judgement that is so swiftly cast through the gaze.

Cyberspace and cybertopias

Although it is not possible to isolate virtual space from the discursive practices of the real, there is an underlying assumption that such worlds offer an escape from the corporeal, such as the technological liberation discussed by Haraway (1991) and Soufoulis (2002). With cyberspace residing somewhere outside our physical realm it is not surprising that there are analogies to a soul-space, to the metaphysical spaces of religious and spiritual significance. Margaret Wertheim explains that such analogies are unsurprising given the associations between immateriality and spirituality: 'It was perhaps inevitable that the appearance of a new immaterial space would precipitate a flood of techno-spiritual dreaming' (1999, p. 257). The search for transcendence online can take the form of organised religion, pseudo-religious practices, or seeking the ideal self. With the ability to produce an idealised simulacrum of oneself, cyberspace offers a utopian space free from physical imperfections and corporeal, bodily 'sin'. It can be seen as a form of paradise where, 'unlike our physical bodies, these cyberspatial simulacra will not age, they will not get sick, they will not get wrinkled or tired' (Wertheim 1999, p. 259). In addition, the places such bodies inhabit can have any imaginable qualities: endless days of sunshine, manicured gardens and temples that invoke age and history, not to mention all the other perfectly embodied souls to interact with. Within virtual worlds there is the potential for experiencing the perfect vision of the afterlife in the current temporal state, almost cheating death in the transcendence of the physical/metaphysical boundary.

Of particular relevance to the analysis of four SL regions, is the symbolism of the classical age, in particular its culture and mythology, as applied to the digital imaginary. In the utopian conceptions of cyberspace, the classical worlds of ancient Athens, ancient Sparta and Eden are viewed as topoi of freedom and perfection. In the collective Western imagination exemplified through mass media forms, the re-creation of the ancient world not only signifies an escape from the everyday but, in the form of virtual platforms, becomes a new beginning, in formation of a world, or worlds, which departs from the history of the corporeal to produce a new idea of the future. The collective imaginary produces its own meanings, which reflect on history not as fact but as a mediated illusion of the past, and thus as what the world can be. This utopian vision is represented most through the spaces that signify ancient Greece with the freedoms and sexual liberties perceived through the lens of contemporary culture. Notions of Eden however, illustrate the tensions in imaging a perfect world and the anxieties with which we conceive new spaces. Concerns regarding gender and sexual conduct are expressed through various historical narratives which the spaces represent; such spaces are an extension of ancient mythologies indicative of contentions over our moral lives. Margaret Wertheim's analogy of cyberspace and the heavens describes the spaces as potential stages for transcendent and liminal experience whereby an individual can undergo a journey of authenticity. This metaphysical conception is useful in viewing virtual platforms as desired utopian spaces. However, the actuality of this can be questioned because of the commodification of the SL platform, as well as the racial, and gendered norms seen within the spaces.

The spaces of Apollo, Eden, Greek Gold and Zeus signify ancient Western civilisation as characterised by the classical world of Greek civilisation and the Roman Empire, as well as the Judaeo-Christian culture, beliefs and religion that followed in the demise of the Greeks and Romans. In the Western historical tradition of Eurocentrism, these cultures have been given primary consideration and significance in contemporary scholastic thought and practice. The ancient world is also held in great esteem in the wider cultural imagination, as evidenced by numerous movies such as *Troy* (2004) and *300* (2006), TV series such as *Xena* (1995–2001) and *Spartacus* (2010–2013), and other forms of popular entertainment such as *World of Warcraft*. This ubiquity has influenced the replication in the digital realm of signifiers of this age. Thus, the influence of the perceived cultural norms of the ancients in terms of heroism, norms of interaction, clothing and costume, sexual conduct, and associated imagery, should not be discounted in the comprehension of the age's popularity. In the contemporary imagination, myth, wish and history blur; it becomes plausible that gods and mythical creatures existed, as the ancient world becomes elevated to occupy a metaphysical realm. Maria Beatrice Bittarello (2008) argues that such conceptions are already 'virtual' in the sense that 'virtual worlds have always existed in literature, religion, and art', whereby the visual and the imagination are the essential components of 'virtual' world construction. She suggests that these components 'reflect and are based upon different readings of philosophical problems pertaining to the definition of what is actual and what is virtual, of what is original and what is a copy (simulacrum)' (Bittarello 2008, p. 2). From this perspective, the ancient world is already, to certain degrees, 'virtual' in the sense of the contemporary popular imagination: it is a place, in the metaphysical sense, of rich imagery, imagined and re-imagined through a complex intertwining of historical, mythical and popular narratives.

More important than any factual account of history are the imagined spaces of meaning. On the platform of SL these spaces acquire a mythological quality, reflected in the island mentality of severance from the 'real'. In addition, such spaces have a transcendent quality and promise immortality. Bittarello (2008, pp. 2–3) remarks that virtual worlds share elements of narrative and promise with historical myth:

> According to the *Genesis* book, the garden [of Eden] is located in the east ... Like Dilmun, this place is connected to the promise of immortality; once Adam and Eve have been exiled, cherubs will guard the road that leads to the Garden and to the tree of life. (3.24) The *Bible* also describes the New Jerusalem, a utopian place, where the lion and the lamb live peacefully together. Not too different are Greek and Roman descriptions of the Golden Age, when 'men' lived as gods, knowing no misery, pain, death.

Narratives within the Bible, as well as ancient mythological stories, extend human attributes to a transcendent form of being that is closer to Heaven: god-like, immortal, abolishing the constraints of the corporeal self. Although more often thought of in the context of classical societies, ancient myths continue to resonate with desires to escape the mundane and the ordinary of Western existence. Through the screen, one's avatar becomes a god-like character which frees the bounded self; the platform is a conduit for the mythical journey that takes one closer to perfection. It is perhaps more accurate to conceive of the adoption of the ancient imaginary in its association with myth. Bittarello asserts (2008, p. 6):

> Scholars studying classical literary texts have outlined the main features of mythic countries and spaces, as described in religious and literary texts ... In the first place, mythic space is intrinsically different from that of everyday life, because it is either inhabited by monstrous or fantastic creatures or because prodigies happen there, or because people's customs are different. Secondly, it is located far away, often to the farthest limits of the earth, so that it can be reached after a long and usually perilous journey – or by using unusual means (e.g. flying). In sum, mythic space is the opposite or reversal of the real world – it can be portrayed as a utopia (Paradise), or a dystopia (Hell).

Using Bittarello's description of mythical texts, it is possible to place representational spaces of antiquity within this paradigm.

One of the great attractions of the ancient world is its appeal as a type of paradise that exists outside of the constraints of the contemporary work-life boundaries. Similarly, Eden connotes a place of relaxation, plenitude and the absence of work. Virtual paradise is a form of 'willed transformation' that is brought about through 'technological transformation, in which a new kind of life has been made possible' through technical innovation (Moylan 1986, pp. 32–33). Extending this idea of Tom Moylan, the SL platform can be seen as a form of literary text where the present is critiqued in reference to an idealised future space. This future, however, invokes an ancient world. In what can be read as a response to the safe, yet highly regulated, conditions of Western society, the ancient world provides escape into a realm where the rules of the present do not necessarily apply. The popularity of games such as *World of Warcraft* and *Myst*, and movies such as *Troy* (2004) and *300* (2006), and TV shows like *Xena* (1995–2001) and *Rome* (2005–2007) attest to a willingness to enter a world that is exciting and action-driven, dangerous and visceral: qualities that have been removed from contemporary Western culture to a large extent. In figuring SL as an extension of this appeal, one discerns less the appeal of action, war and blood (although there are role-playing areas for combat within the platform) and more the promise of transcendent engagement seen in Apollo, Eden, Greek Gold and Zeus. SL offers the ability to define one's own terms of engagement, interaction and behaviour, within the choice and scope of each region: its attraction lies in the leaving of the 'everyday' to experience the desired. As is evident within the four regions studied, such desire can be evoked in the representation of the self as desired, or as celebrated in difference, as well as in the potential for identity experimentation that may not be permissible at 'home'. As areas explicitly demarcated as sites of 'difference', they appeal to diverse modes of sexuality experienced through an imagined ancient culture. By entering into the ancient virtual space, one is transported to a metaphysical ancient space of re-imagined history, mythology and culture.

Of particular appeal to the contemporary Western imagination is the sexual freedom permitted by ancient civilisations: ancient Greece is perceived as being synonymous with male homosexuality and pederasty; ancient Sparta evokes the independence, freedom and strength of women. Similarly, the Judaeo-Christian conception of paradise, the Garden of Eden, is synonymous with freedom and temptation. It is the *idea* of these civilisations, rather than any historical actuality, which provides the impetus for exploration of the self, sexuality and identity. The three major platforms for inquiry (ancient Greece, ancient Sparta and the Garden of Eden) not only share imagery of nature and an Epicurean ethos, but importantly evoke an ease with sex and sexuality and its celebration (although the end of the Eden narrative is most often read as a tale cautioning against such celebration). In a contemporary Western world fixated on risk avoidance and conformity, the cultures of hedonism and free sexuality as experienced by the mass public are rare, although it must be noted that these behaviours are widespread in various forms of media and entertainment (conformity as the everyday, subversion as spectacle). In reference to Judith Butler, idealised spaces are constructed in view of the established fields of normative sexual practice; they do not reveal a purer, historical form of sexuality, but rather reveal an argument against, and critique of, our established conditions of sexual practice. Such spaces do not hark back to a nostalgic past, but rather imagine a desired future. In the age of digitalism one might expect to find region after region boldly expressing new visions for the future, perhaps reflecting science fiction ideas of high-tech cities, shiny metals, robots and alien landscapes. Although such virtual spaces do exist, the desire to re-create, albeit idealised versions of, ancient times is evident not only in the SL spaces researched, but also in movies, books, TV series as well as video games and other 3D virtual environments. Such re-creation may not simply be an imagining and idealising of the past, but may also reflect desires for the future. The search for authenticity can be attempted through the liminal appeal of the platform: a search for an authentic centre with other like-minded individuals in a collective exchange of imagining. As stated by Margaret Wertheim, 'The promise of utopian community is one of the primary appeals of cyberspace' (1999, p. 283). In a world suffering from economic disparity, alienation and disaffection, it is unsurprising that individuals may look for a utopian vision elsewhere.

Notions of widespread change, however, are premature in consideration of the encroachment of CL regulation and policing of the virtual domain. Freedoms are becoming limited in the moral panics of wider society regarding the behaviour reported in various sites. SL has been a particular target of moral and ethical panics about in-world conduct since its inception. Although no evidence was found of pederast or paedophilic conduct on Apollo or Zeus during the study, there is reportedly a subculture in SL involving 'age-play', 'child-play' or 'edgeplay' characterised by 'residents embodying themselves as children and having sex with adult avatars' (Linden Lab Official: Clarification of policy disallowing ageplay 2011c). This has prompted scrutiny of the environment, eliciting fears that tap 'into fundamental questions of agency, power, and propriety with regard to sexuality' (Boellstorff 2008, p. 164). As a response to such concerns, Linden Lab has clarified its policy on 'age-play' within SL (Linden Lab Official: Clarification of policy disallowing ageplay 2011):

> Under our Community Standards policy, real-life images, avatar portrayals, and other depictions of sexual or lewd acts involving or appearing to involve children or minors are not allowed within Second Life. When detected, individuals and groups promoting or providing such content and activities will be subject to sanctions, which may include termination of accounts, closure of groups, removal of content, and loss of land or access to land.

Such concerns over behaviour illustrate the on-going tension between the corporeal and virtual worlds and the apparent need to regulate all facets of social behaviour, particularly in reference to sex and sexuality. Whilst many users participate in SL to escape from the constraints of their corporeal existence, it seems that the popularity of such worlds may predicate their continuing regulation to the point of indiscernible difference from the 'real'. Furthermore, such debates illuminate the permeability of the line between the corporeal world and 3D virtual spaces; to what extent any behaviour is 'real' is in question, including the potential repercussions and consequences of such behaviours. Although outside the scope of this text, such questions will continually inform the culture of SL spaces and virtual worlds at large. What this issue greatly demonstrates, however, is the encroachment of real-world norms onto virtual spaces and the freedoms they represent. The divisive nature and judgement of 'age-play' as a practice in-world, the reactions to it by the Western media, and the response by Linden Lab show the pervasion of corporeal social standards into every facet of life, including the virtual, the representational and the imagined.

Moreover, and particularly in consideration of SL, is the duplication of capitalist practices of CL in the virtual. The presence of the Linden economy has had similar effects on the virtual residents in SL as the Western globalised economy has had on nations and communities in the corporeal world. The Linden economy is based on the US dollar and user-generated content, allowing 'users to retain property rights and, as an effect, profits on the things they create' (Castronova 2005, p. 167). Whilst there may be an egalitarian ethos in terms of the identity one adopts, this is still constrained by one's capacity to buy goods and services in-world. A class system exists in which those who have the capital are able to represent themselves in a more desirable fashion, as well as displaying conspicuous consumer traits such as owning land, houses and leisure goods. In addition, one is able to animate one's avatar to suit, as well as purchasing more realistic skins and avatar parts. All these elements work together to produce a class system in-world that differentiates those who have mastery and agency from those who do not. Furthermore, there is a distinct racial dominance of whiteness on the four sites visited which illustrates the racial divide in SL as well as cyberspace (Nakamura 2008).[1] Remarking on inequities on the internet, Wertheim suggests that we are seeing a replication of colonialism in terms of Western, white, male affluence (1999, pp. 296–297). This is certainly the case in the spaces that were surveyed, with the notable exception being Greek Gold in its representation of diverse gendered displays. Even then, the avatars encountered were all white, and notably youthful. Homi Bhahba (1998, p. 21) affirmed in his critique of whiteness that the normative position in Western society is the white position and that this is a strategy of authority. As such, the pervasiveness of white avatars reproduces a position of social and cultural agency that is privileged and powerful. As SL tourists, performing whiteness plays out white-neo-colonial fantasies of privilege and esteem.

Conclusion

Technology and the digitisation of experience extends both the possibilities of embodiment as well as reifying the body in reference to its static corporeal conceptualisation. It is perhaps this latter reason that feminist and queer scholars use as a foundation for re-thinking theory and politics, and it is certainly this motivation that drives the investigation into the four regions of Apollo, Zeus, Greek Gold and Eden. In extending our reality into the virtual realm, can we really re-think our bodies and the spaces they inhabit, and perhaps even transcend them? Whilst initially cynical about such grandiose notions, lived experience through an SL avatar both illuminates and extends the static imagining of the body and space as natural, as closed, and as fixed in its meanings. Even in imagining and creating utopian spaces, the spaces and our bodies within these spaces are written with prescribed meaning. It is not as if we can start with a blank slate and envisage cultures completely detached from our physical present – every space becomes an extension of a pre-existing one. Imagined space speaks more of our desired future than an idealised past. It speaks of a yearning to escape the present, yet can only extend itself in a limited fashion as it is forever anchored to that which it desires to escape from. In the following chapter these notions of utopia and escapism are framed through tourist discourse, which expands the ideas of historicity, tradition and authenticity in the virtual environment.

2
Tourism: Island Utopias in the Virtual Sun

Introduction

Second Life is highly representative of tourism in that an individual embarks on both escapism and fantasy from their CL. In evoking John Urry's work on the tourist gaze, this chapter is not only an analysis of SL spaces through Urry's defining terms, but also an extension of a discourse about a physical phenomenon to the virtual. The virtual environment of SL is similar in function to CL tourist spaces due to the evocation of the gaze and the transportation to another environment, outside the realms of work and domesticity. Urry contends that tourism 'necessarily involves some movement through space, that is the journeys, and periods of stay in a new place or places' (2002, p. 12), and that places are chosen because of 'intense pleasures, either on a different scale or involving different senses from those customarily encountered' (2002, p. 3). So, although there are obvious corporeal differences in how the space is experienced, virtual environments evoke many of the same phenomena of CL travel and tourism. Moreover, Urry and Larsen (2011, p. 2) contend that '[G]azing at particular sights is conditioned by personal experiences and memories and framed by rules and styles, as well as by circulating images and texts of this and other places.' This is particularly important in the consideration of virtual space and the foundations of looking that we bring to it, as the images and texts that signify tourist sites are remembered from our own photographs of travel as well as well-known texts and images of popular sites.

Using tourism as a discursive framework, this chapter provides not only a general analysis of the virtual as a tourist experience, but also details some of the specific phenomena of the tourist experience. The ideals of coastal travel are discussed through the symbolism of the Mediterranean as apparent in all the spaces looked at for this text, after which they are analysed utilising Gray's (1970) motivations of 'wanderlust' and 'sunlust'. Firstly, Apollo and Greek Gold are framed as spaces of wanderlust, specifically assessing notions of authenticity and inauthenticity which draw upon the contradictions of history and mythology, as well as the idea of liminal experience through ritualised engagement and participation in the mythological narrative of the spaces. In evoking the Mediterranean and various 'natural' and 'built' environments, the spaces show a desire for aesthetic experience contained within the construct of authentic experience. Using Kant's notion of the sublime, ideas of human mastery over nature are discussed, as well as the tension between the pleasure and displeasure of the virtual environment. Turning from ideas of nature to the body is the concept of virtually embodied pleasure. In contrast to the ideals of authenticity, Eden and Zeus are analysed as sites of pleasure centred on the body, including bodily performance, hedonism and liminoid (liminal without ritual) experience. Explicitly delineated as gay male and gay female spaces, they are discussed in reference to queer tourism, a burgeoning arm of the leisure industry that has particular ramifications for identity as well as bodily practice.

Viewing virtual worlds as the margins of 'everyday' work and domesticated experience, they can be conceptualised as the new 'pleasure periphery' of sex, sun, sand and surf (Selanniemi 2003). Louis Turner's and John Ash's tome (1975), which describes the post-war shift in leisure experience to include the masses, can be similarly utilised to describe the shift in experience from the corporeal to the virtual. The body is experienced as differently embodied on and through the screen. Similarly, SL leisure, pleasure and hedonism are changed to incorporate a different sensory experience from 'real' travel, although still concentrated on the body as the site of meaning and desire. In the postmodern globalised world, where the leisure industry has grown to include all continents and every conceivable experience, the new periphery can be found through the screen as adventure and escapism that transcend the common experience of physical travel. Within this shift, however, is found a reiteration of the political fabulations of the physical tourist experience. Principally, these experiences are grounded in the pleasures of the male gaze and patriarchal agency which permeate sites of queer experience through representational practice. This has particular ramifications for queer sites, specifically those centred on female engagement, displaying a marginalisation of effective agency applied to spaces that should transcend this power structure.

The (virtual) tourist gaze

SL is a world that is consumed, most often in a visual sense, and most often for pleasure. As stated by Urry, the act of looking is formulated upon difference, 'in relation to its opposite, to non-tourist forms of social experience and consciousness' (2002, p. 1). As a platform intrinsically different from the realms of work, domesticity and ordinariness, SL embodies difference both in form and function. As an escapist environment, SL evokes the ideals of travel and tourism. It is a space where people can escape the everyday and 'travel' to experience scenery, serenity and beauty, and, importantly, deviance and the departure from acceptable social practices. Central to the tourist experience is symbolic capital, or the reduction of experience to sign value – one escapes the everyday through the consumption of tourist and leisure signification which is the site to which pleasure and deviance are attributed. Due to the high dependence on the visual in SL, this symbolic capital is both essential and necessary to the experience of departure and immersion. If one wishes to escape from CL and its mundanities, one expects visual stimulation and excitement over and above that experienced in CL. Characteristic of the postmodern emphasis on visual culture, much of the symbolism within SL is of simulacra, a plethora of signifiers joined together to create a hyper-real experience. The symbolic capital of the studied regions has no essential referents, instead producing a multiplicity of meanings and often drawing upon symbols that have a tenuous place in history and culture.

The regions are spaces where residents and users of SL 'travel' for even greater escapism. There is an irony in SL users travelling to Apollo, for instance, from other SL regions and residences as a destination that evokes the tourist experience in a platform that is inherently escapist. Although SL is often assumed to be a place of departure from the real world of work and domesticity, there are also places of work and home-life in SL, particularly as some users spend more time in-world and use the platform as a source of income. Whilst outside the scope of this analysis, such spaces can be assumed to operate somewhat similarly to work/home spaces in CL, with comparable delineations of the functioning of space dependent on the activity performed in that space. Therefore, for a SL resident who spends much of their time selling real-estate in SL, their work space could be confined to the islands which they are building, potentially spending time manipulating 'prims' (the building blocks of SL construction that all objects are made from) and promoting their estate/s. Unlike much of SL, Apollo, Eden and Greek

Gold do not house residents or land-owners, so avatars cannot 'live' on the islands. Zeus, however, has a residential estate that is connected to the entertainment precinct.

The underlying purpose of these SL regions, though, is for seeing and experiencing. It is this foundation of travel and visual experience that sets these islands apart from other SL regions in that they are structured for the tourist gaze. Practices of looking are privileged in the tourist experience, and experience in general, where 'sight has been regarded as the noblest of senses' (Crawshaw & Urry 1997, p. 177), so that 'sight becomes highly significant in the ordering of tourist and travel discourses' (p. 178). It is unsurprising that virtual experience is dependent on the visual as the site of experience and, similar to tourism and travel practices, invites pleasure through looking. This can be either through the solitary romantic gaze at the extraordinary, or the collective liminal gaze for looking at others and being looked upon.

As a term initially used by Turner and Ash (1975) to describe the regions of the world that were used after the Second World War as places of escape for tourists from rich Western countries, the 'pleasure periphery' can be expanded to encompass mass tourist practices focused on the pleasures of the body. Centred on sun and sex, these destinations were, and still are, utilised as escapes from the everyday for hedonism and indulgence. It is now no longer necessary to physically travel to find such escapism. Virtual worlds can be considered the new pleasure periphery in which users depart from their everyday in search of sun, sex and hedonism akin to physical beach resorts. In fact, it may be possible for individuals to experience an even greater focus on individualised pleasures given the ease with which one travels virtually through differently themed spaces. Behaviour within these spaces can be described as liminoid, or 'those socially accepted and approved activities which seem to deny or ignore the legitimacy of the institutionalised statuses, roles, norms, values, and rules of everyday, "ordinary life"' (Lett 1983, p. 45). Liminoid behaviour is often characterised by tourist behaviour and experience that, to some degree, inverts the behaviour and experience of the everyday insofar as the tourist space introduces a change in the customary behaviour of the individual and both allows and encourages other forms of identity performance and sexual practice. Or, as described by Selanniemi, 'the liminoid destination encourages behaviour that would not normally occur at home' (2003, p. 25). Sites specifically based on hedonism, sexuality and sexual expression offer readily accessible zones for instant bodily gratification and interactions, including the possibility of sexual liaisons. Such interactions are heavily reliant on bodily appearance in-world and, like the physical pleasure periphery of CL coastal resorts, are moderated, and to a large degree controlled, by the male gaze.

Evoking the Mediterranean

In producing idealised visual environments, the spaces all utilise symbolism that evokes the Mediterranean. White beaches, perfect blue skies, rocky cliffs and dramatic island mountains and coastlines are present in all of them. Interspersed with these 'natural' features of the landscape is the built environment of romantic enclaves, historical symbols, and structures of spectacle that dominate the 'nature' below. These natural features are more prevalent on Apollo and Greek Gold than on Eden and Zeus, although Zeus has these features in its residential zones. Whilst Apollo draws upon historical symbolism, in terms of its evocation of ancient Greece, Greek Gold, Eden and Zeus suggest a more contemporary Mediterranean with their nightclubs and modern structures surrounded by the 'natural' landscape of beaches and mountains (this is very limited on Eden). Interestingly, Eden's snow-capped mountains are entirely two-dimensional, without the slightest suggestion of trying to replicate anything like 'real' mountains. Instead, they act almost like a shield, protecting a side of the region from unwanted gazes from elsewhere whilst providing a tourist-view/backdrop, however paradoxical to its beach-side setting. Although there is a binary between images of the snow and beach, these sites are evocative of idealised touristic destinations, which convey a leisurely escape from the 'everyday' real world. Such significations, with varying degrees of mimesis, evoke nature and, more specifically, the frontiers of adventure. Their evocation of the Mediterranean, as well as European mountainscapes, suggest an attachment to the Grand Tour of the seventeenth and eighteenth centuries. Nature as conceived on the Tour, which continues through to today's relationship with the natural environment, was experienced through subjective pleasure in looking. Joseph Clark (2011, p. 153) points out that this ideological framing of nature even permeates the locals' perspectives in viewing the environment as visually consumable.

A comparison between images of the virtual environment of SL and those advertising the Mediterranean reveals a consistency. Synchronous with the 'real life' pictures of the Mediterranean, their appearance captures the island fantasy of idyllic escapism. This synchronicity exposes the common imagery of the Mediterranean as simulacra; the form within the images of 'the real' and SL are almost interchangeable, and both ultimately a reference to the imaginary. The coastline not only encapsulates ideas of beaches and fun, but also the ancient mythology of the Mediterranean, representing 'the fringe between the known and the, often frighteningly, unknown' (Ryan 2002, p. 156). For Apollo, as a self-contained space, the island fantasy evokes even more in terms of adventure and mystery. For the other regions, however, the coast symbolises hedonism. The tourist gaze is elicited whereby the individual is invited to look upon the scene and incorporate the self as part of the historical imaginary, as well as looking at others and being looked upon. The regions of Greek Gold (an island), Zeus (an island) and Eden (secluded from the mainland) achieve separation and security from those who are not permitted to gaze upon them – it evokes 'islandness' by not being at the margin, but separate from it. (Sharpley 2004, p. 23). Seclusion and isolation are achieved differently in the locations. However, all represent 'an escape wherein the person can attempt to find themselves' (Ryan 2002, p. 157). To complement the feeling of separation from CL, there is the caveat of gateways of participation. Apollo maintains a sense of sparsity through the capabilities of the software; only a certain number of avatars can visit the island at any one time, otherwise 'rezzing' would be rife (see Chapter 3 which discusses the island of Apollo at length). Eden and Greek Gold, on the other hand, maintain their exclusivity through gender (only female avatars are allowed), and Zeus through the separation of the space from the rest of the land it is attached to as well as being marketed to the gay male demographic. This isolation is critical to a sense of escape: both geographically and spatially, from one's home and typical surroundings, and from the typical interactions and spatial configurations found in Western cities and suburbia. That the island and coastline are so contrasted with the everyday is paramount to the imaginary; the imaginary must be what the ordinary and everyday are not. As a space of exploration and possibility, the imaginary must project an environment and spaces that encourage different ways of being, particularly in reference to hedonism, romance, sexuality and narratives of adventure (historicity and mythology).

According to Sharpley, the features of islands, namely 'the sense of distance, exclusivity, separateness, insularity [and] tradition' (2004, p. 23) can be termed an 'islandness' that embodies a sense of play and fascination not found elsewhere in society, particularly Western society. This 'islandness' elicits a type of behaviour and evokes certain cultural memories that contrast with contemporary life and social expectations. Unlike the rigid spaces of home, work and family, islands allow for an exploration of behaviour: 'They are places to which tourists are often attracted by the perceived opportunities for hedonistic or unlicensed playfulness, a fantasy "other" affording opportunities for activities or behaviour constrained by the customs and structures of everyday home life' (p. 23). The four regions are spaces that represent "islandness" through their appeal to island symbolism and their explicit permission of subversive behaviours. This 'islandness' is maintained principally in the following ways: the visual imagery of the island itself (its form, structure and environment), behavioural expectations and guidelines (including normative group behaviours), and the cultural expectations associated with a place defined by such visual and behavioural cues. Even in SL itself, there are places designated for certain activities, so by explicitly defining a space through symbolism and behavioural expectations, space creators can define areas specifically as island playgrounds. Not all islands in SL are for such a purpose, as many areas are in island form simply due to their ability to be discrete and separate from other SL spaces, therefore maintaining a sense of exclusivity for purpose (educational and work spaces for instance).

The idea of the island as utopia has carried through modernity, representing an idealised frontier. Ever present is the idea of escape and its opposite, restraint, through which utopia is conceived as:

> on the one hand, a free play of imagination in its indefinite expansion measured only by the desire, itself infinite, of happiness in a space where the moving frontiers of its philosophical and political fictions would be traced; on the other hand, the exactly closed totality rigorously coded by all the constraints and obligations of the law binding and closing a place with insuperable frontiers that would guarantee its harmonious functioning. (Marin 1993, pp. 403–404)

Here, Louis Marin acknowledges Thomas More's utopia as the dialectic coupling of both the restrained imagination and the hope of emancipation as they relate to the lived world in here (represented by the civilised city-state) and the frontier world out there (represented by the absence of civilisation and the built environment). First appearing as the title of Thomas More's book, later applied to 'the name of the island described in book two of *Utopia*' (p. 407), the word is also a 'Latin neologism from a fictitious Greek word: *ou-topia* or *eu-topia*'. More's conception has carried over into the contemporary imagination whereby the Greek island, especially in relation to European hedonism, has become synonymous with the word. The island is particularly important in the imagination of utopia because it signifies a boundary between the two worlds (of civilisation and the frontier):

> Utopia is the figure of the horizon. If in the functioning of a city, in its structure formed by streets and dwellings, and if in the functioning of a landscape, in its partition between nature and culture, forests and fields, waters and rocks, space cannot exist without limits and frontiers, Utopia as a city or a landscape develops and displays a virtual or potential spatial order in its text, it offers to the beholder-reader an ambiguous representation, the equivocal image of significations contrary to the concept of 'limit': on the one hand the synthetic unity of the same and the other, of past and future, of this world and the beyond (and the frontier would be in this case the place where conflicting forces are reconciled), and on the other hand the active tracing of differences, the indefinite fight between opposite forces (in this case the frontier would open a gap, a space 'in between' that could not exist except by the encountering of violent and resisting forces). (p. 412)

It is the dialectical arrangement that defines both worlds in relation to each other. The civilised built environment of the Western city-state is replete with certainty, monotony and reason, which is opposed to the frontier of the utopic environment with its nature, ambiguity and spirituality. Utopia is the desired place of transcendence, although fleeting in its presence owing to the nature of the dialectic relationship. Ultimately, utopia reveals itself primarily as a psychological state – its brief temporality holds transcendence as a transient phenomenon catalysed by some bodily sensation of presence.

The frontier in the postmodern world has shifted from the tangible period of exploration and imperialist conquest of cyberspace and the infinite possibility of space itself (Marin 1993, p. 406). The SL spaces represent two overarching motivations of tourist consumption that seek different forms of utopia: that which centres on the body and that which seeks transcendence or authenticity. These two categorisations draw largely upon Gray's motivations, defined as 'sunlust' and 'wanderlust' (1970). As elaborated by Markwick, 'sunlust' 'may be more appropriately reconceptualised as the ever more extensive and intensive search for the exoticism of "the pleasure periphery" (rather than merely a search for better amenities and climate elsewhere)' (2001, p. 421). 'Wanderlust', however, can be categorised as a search for the authentic, in terms of ritual, the sacred and myth that often evoke the romantic through historicism. It is the search for the 'other' in terms of travel from the domestic space of alienation in search of some authenticity grounded in history, culture and myth. Central to the utopic journey is the search for authenticity as one seeks out experience and space that are closer to some truth than those of the lived world back there.

The (in)authentic

Of the four regions, two can be considered to provide an experience of authenticity, albeit for different reasons. Authenticity, as applied to travel and tourism, has two key interpretations. The first is in 'MacCannell's concept of tourism as the pursuit of staged authenticity', in a search for history and culture that is unspoilt (Brown 1996, p. 37). Apollo can be considered a destination that offers this sort of authenticity in that it appeals to historicity and mythology through its environment (ancient Greek structures) and behavioural codes (having mostly single to small groups of avatars rather than large groups of avatars as found on other regions). However, and as argued by Hillman, this does not conform to the more recent conception of authentic tourism 'in the sense of an unaltered product' (2007, p. 5) that posits authenticity as external to the individual. Moreover, as argued by Brown (1996), the search and ultimate transformation of the space due to the search for the authentic and unspoilt gradually transforms it into a consumable destination of the masses. Alternatively, authenticity in travel can denote the journey to find one's true self. Described as liminality, this search seeks a transcendence catalysed through discovery of place that elicits a new knowledge of self in what could be considered paradise. Greek Gold appeals to the search for the authentic within the self by offering an environment of tolerance and acceptance that allows for the expression of a desired self.

As elaborated in Chapter 3, Apollo draws upon ancient Greek mythology for its structural artefacts and environment, but also, and very significantly, for its symbolism of gender and sexuality. Unlike Eden, a lot of detail has gone into the production of Apollo as a place that extends symbolism into a form of authentic experience. This experience relies heavily on cultures of history and mythology to evoke the authentic as offered through ancient Greek signification and symbolism in their most mass produced and recognisable forms. Mass production of form is obviously antithetical to the authentic, but this is not just a paradox of authenticity-seeking in the virtual realm: it can be argued that any tourism, whether virtual or corporeal, ultimately seeks to consume ideas of history and culture in 'staged authenticity'. Therefore, as Harkin notes (1995, p. 652), the tourist experience, especially of and by the West, is centred on a hyper-reality of the visual and the simulacra of form, where 'authenticity and inauthenticity (the genuine and the fake) feed off each other in dialectical fashion' (Brown 1996, p. 37). Indeed it may be that the virtual is a more authentic experience of the authentically hyper-real, in that it dissolves any misapprehension about authenticity based in actual truth, in that this dissolution of truth (or the ultimate revelation that our lives are not based on any notion of objective truth) is the most authentic experience possible when history and culture are accounted for.

Potentially, what is more important in terms of authenticity is not the space itself as a true marker of 'real' experience, but the idea of lived experience (phenomenological) as precipitated by signification. That the various cultural and historical markers are very obviously not physically real is no hindrance to what they represent in their implications for ways of being and experiencing. On Apollo, it is the signification of the historical that implies the myth and the narrative of that space. This ultimately allows for the individual to experience the space as part of that narrative, rather than simply viewing a place of history; the very virtual-ness of the experience allows for more direct participation into the imaginary of the space, and the history and mythology it represents. Rather than simply being part of the scenery or being static in view of the landscape, one may participate with others in acting out what it means to be part of the place. From this perspective it can be argued that spaces like Apollo are a different type of tourist experience, in that they elicit an experience akin to participation in the imagined place of historicity. So, rather than simply evoking the historical gaze, Apollo encourages participation as part of the narrative of history. Furthermore, and as explained by Joseph Clark, the choice to re-create natural and built features that signify the corporeal demonstrates an unshakeable attachment to the physical, historical and cultural world – we do not necessarily wish to fulfil a post-human future of detachment (2011, pp. 145–146).

Participation in the narrative, similar to the ritualised journey of the tourist, is part of the escape from the domestic and work spaces of the corporeal environment. This concept is described by Harkin (1995, p. 656):

> The specific content of the exotic may vary, but it always represents an alternative to domestic experience and discourse. The exotic is

> highly cathected, especially as a category of psychosexual otherness. The potential proliferation of new social meanings created by the exotic is dialectically opposed by domestic ideological structures, structures that play an important role in tourism.

The exotic in the virtual, prompted by visual signification, has the same effect as the exotic in corporeal tourist practices and evokes behaviours assumed by the exotic, having a phenomenological effect. Experience becomes exoticised, as part of the dialectic of otherness which the self engages in, and often becomes a part of. Similarly, the exotic is part of the practice of cultural differentiation whereby the self may discover authenticity through otherness. As described by Harkin (1995, p. 666), this process is similar to anthropology in that a culture is reified in comparison with others. Although mostly used in theory to discover and compare discrete and whole cultures and societies, it is possible to appropriate the concept of authenticity in the analysis of the self-subject, where cultures are seen not as discrete units but heterogeneous and fluid. Here, authenticity is located at the site of the individual, where exposure to a different culture and mythology, and the participation in a narrative of otherness, can precede a knowledge of self similar to liminal engagement. In a phenomenological sense, the experience of participation within the narrative allows for a cognitive shift in the self, and is reflected through experiences of, and about, the body.

In addition to the exoticisation of culture is the romanticisation of the virtual landscape. As described by Barthes, the desire for the picturesque has been held as an ideal bourgeois reason for travel, epitomised in the Grand Tour of the British aristocracy. He explains (2000, p. 74):

> We find again here this bourgeois promoting of the mountains, this old Alpine myth (since it dates back to the nineteenth century) which Gide rightly associated with the Helvetico-Protestant morality and which has always functioned as a hybrid compound of the cult of nature and of Puritanism (regeneration through clean air, moral ideas at the sight of mountain-tops, summit climbing as civic virtue, etc.).

Of the four spaces, Apollo more than the others represents landscapes of untamed nature, offering the most 'realistic' imagery of a mountain wilderness, or depictions of the sublime. Forming part of the search for authenticity, Kant's notion of the sublime describes not only our relationship with nature but also, and most importantly, our superiority over it (in later appropriations of the theory). The produced virtual environment iterates the idea of human mastery over the natural environment and our feelings of power and rationality over nature. The virtual environment illustrates Kant's idea of the sublime as existing only within the mind, and not within objects themselves, as well as questioning the dominion of nature. Using Kant's notion of 'negative liking' as applied to the virtual landscape, the spaces show that we can experience a 'mixture of pain and pleasure' (Guyer 2006, p. 322), feeling pleasure in nature's replication but also displeasure in that the nature created is not 'real' and therefore has no dominion. The dynamical sublime that we experience here is slightly different from that evoked by the natural environment, displacing 'the element of fear and pain at the thought of our own physical injury or destruction' (Guyer 2006, p. 322) where we ultimately feel powerful in the rationality of the threat, and in the idea of power we feel at mastering the threat of nature in that we have become its ultimate creator. The idea of creating, witnessing and playing within the extraordinary enhances our illusion of superiority whilst exposing us to the extraordinary. It can be argued that the (CL) mountains and wilderness, indeed all landscapes of the sublime, have become banal to the Western imagination by being so accessible and widely visited. Elements of mystique have all but disappeared due to mass tourism, and the feeling of initial inferiority described in Kant's dynamical sublime has largely been surpassed through our physical mastery of the environment (even the most inhospitable places have become tourist attractions in recent times). Chris Rojek argues that 'the evaporation of physical distance eliminates the sense of remoteness which is the essential requirement in perpetuating the authority of the auratic object (because) through ease of contact via mass reproduction, the object becomes just another found object in the showplace of mass culture' (1997, p. 59). In this erosion of the extraordinary in terms of travel as predicated by mass culture, extraordinary experience is sought elsewhere. New spaces of exploration are sought, as are new means of travel. Virtual travel, whilst still reliant on the invocation of history and mythology of the culture from which it is derived, allows for participation in newly formed narratives of the extraordinary.

The type of space visited elicits a different form of extraordinary experience from the everyday, whether this is as part of a liminal/transcendent (Apollo), or liminoid/hedonistic (Eden) journey, and these differences depend very much on the structure, as well as allowed behaviours and actions, of the space itself. This difference in form imparts a very different experience. Due to the capacity for participation allowed within the structure of the spaces, Apollo is a wanderlust destination, whereas Eden is principally structured around experiences of the body. The space of Eden is framed by romantic imagery, evoked at a distance, a mere reflection of possibility rather than participation. Apollo, on the other hand, invites participation as part of a romantic/Romantic narrative. Apollo, as completely separate from other SL regions, invites individuals to be part of the ritual of finding themselves, and of reflection, through the capability of embodied movement. It is what Nelson Graburn (1983) would describe as a ritualised tourist experience, or an invitation for individuals to partake in an experience of potential transcendence of the self. Through the adoption of another, or many other, role/s the individual is free to explore and to participate in the (new) mythology of the space; on Apollo, this could potentially be any identity aligned with the classical period of beauty, sexuality and hedonism (although, as described in the next chapter, such identities would have to be in keeping with normative representation).

A potential (and obvious) criticism, however, is that virtual spaces are implicitly inauthentic due to their existence outside of our normative corporeal environment and senses, and therefore cannot deliver any 'truth' about our corporeal reality; that, due to the visual replication of signifiers, places in SL can no more deliver an authentic experience than can a popular text drawing upon mass produced imagery of a mythologised past. However, such inauthenticity is inherent to all conceptions of history and, as in the case of Apollo, a romantic mythologisation of the classical world that, in essence, can never really be uncovered. As detailed in Chapter 3, Apollo evokes much of ancient Greece in symbolism and discourses of presentation and behaviour. However, it also evokes the tourist gaze through participation in the classical world, as one travels through and interacts with the space. Any access to a possible 'truth' of the classical is removed, leaving authenticity at the site of the individual subject. It is not only ancient Greece that is depicted on Apollo, as there are referents to the Mediterranean as a whole, underlying the space's ability to evoke not only a single culture of the classical world, but spanning various mythologies. For instance, a designated area for dancing on Apollo called Salsa y Boleros (translated loosely as 'dance and music'), signifying Spanish-ness through dance, is situated on a raised platform overlooking an idealised seascape. That elements of Spanish-ness are included on the island of Apollo appears to conflict with the majority of the island's structures which connote ancient Greece. However, it must be understood in the broader context of other-worldliness that the island seeks to evoke. Through a pastiche of signs and signifiers, representations are '"healthier", that is, less ideologically deceptive, when they draw attention to their arbitrariness and do not appear as innocent, unmotivated, eternal or natural' (Duncan & Duncan 1992, p. 19). So, the 'staged authenticity' of tourist practice becomes entirely transparent, revealing itself as an obvious cultural replication.

As an entirely fabricated mythological environment, Apollo can be considered more authentic than many corporeal tourist sites as it deconstructs the illusion of history and uncovers the cultural foundation that makes all experience mediated: perhaps inauthentic to 'truth', but authentic to the self. Barthes argues in *Mythologies* (2000) that travel and tourism have an effect on the sign value of environmental and historical features, a point expanded upon by James and Nancy Duncan who state that there may be an '"overpowering" or even "masking" the "real" spectacle of human life and history and simultaneously providing an illusion of cultural stability and continuity' (1992, p. 20). Becoming a part of a mythology on Apollo, however, is not the same as 'buying effort', which Barthes attributes to the bourgeois sightseer reliant on paying others for means of travel and cultural experience. In contrast to this position, participating in SL consumes much time and effort from a participant, not only to learn the features of the platform, but also in terms of avatar investment; participating in SL is certainly not passive. It is also unlike the 'staged authenticity' of tourist centres and attractions, whereby cultures enact authentic experiences for the tourist, as is often derided by academics and theorists (MacCannell 1999, p. 102):

> The tourist experience that comes out of the tourist setting is based on inauthenticity and as such it is superficial when compared with careful study. It is morally inferior to mere experience. A mere experience may be mystified, but a tourist experience is always mystified. The lie contained in the tourist experience, moreover, presents itself as a truthful revelation as the vehicle that carries the onlooker behind false fronts into reality.

Apollo is not a staged environment in the same way that tourist sites stage experiences for tourists: in that it does not represent the real of history or culture, or 'genuine fakes' as described by Brown (1996). It is similar to Disney and Las Vegas in its obviousness, yet successfully provides escapism through the imagination of the consumer (Gottdiener 1995). It is a fully artificial and fabricated environment that signifies a mythology of the classical period, but not in any historical sense. Rather, it represents a set of behavioural codes and performative evocations through imagery, symbolism and norms. As noted by Cohen, 'travelling for pleasure [as opposed to necessity] beyond the boundaries of one's life-space assumes that there is some experience "out there", which cannot be found within the life-space, and which makes travel worthwhile' (1979, p. 182). Therefore, the authentic may not lie in the destination itself, but in the experience evoked by the journey to and through that space.

It is possible to view this form of authenticity as a type of liminality, a transition over a threshold or journey of transcendence, usually in search for a more authentic 'centre'. Cohen describes the search for a 'centre' as something that came about in archaic times, 'when a powerful mythological imagery locates the "real" centre in another place, beyond the limits of the empirical world, a "paradise" beyond the surrounding chaos' (1979, p. 182). In developing Turner's (1977) concept of liminality as applied to tourism, Graburn (1983, p. 14) emphasises the changes experienced in social and cultural norms, as a common aspect of tourism:

> During this period (i.e. after the preliminaries, while game is in progress, while the tourist is away from home, at leisure away from the workplace) the social and moral structure of the group is changed. The social patterns are different from the normal, sometimes including reversals of roles, in this period which Turner characterizes as antistructure. The content of the social relations is no longer normative, hierarchical and distant, but close and egalitarian, a state that he calls communitas: in communitas, which may be spontaneous, or normative (expected) or even ideological (e.g., forced by a church, government, party, or even by the staff of a holiday camp), symbols of rank and status may be shed and people feel they are communicating with each other as individual persons rather than as performers of roles.

This is similar to what Cohen describes as the 'existential mode' of tourism, in which the individual seeks out immersion into an 'elective centre'. The tourist:

> will live in two worlds: the world of their everyday life, where they follow their practical pursuits, but which for them is devoid of deeper meaning; and the world of their 'elective centre', to which they will depart on periodical pilgrimages to derive spiritual sustenance. (Cohen 1979, p. 190)

The notion of communitas is illustrated on Greek Gold where the prescribed social relations of the space are far removed from the everyday work/domestic sphere of the corporeal. The idea of transcendence in paradise is evoked through the idea that one can be one's true self, stripped of the inauthentic expressions and symbols that may characterise one's performed or unreal self. The decentred, virtual embodied self travels to the space for 'spiritual sustenance' and the reaffirmation of the non-normative self which may be in part lost through the everyday. Not everyone who participates in the SL regions is actively seeking such a liminal experience, however. Experiences that are more 'recreational' (and less transcendent), to use another of Cohen's five modes of tourism, are often centred around the body: interaction, performance and hedonism.

The body

Whether travel's major purpose is purely recreational in nature (less authentic) or liminal (more authentic), the body is central to experience. The body is both the site of pleasure and the centre of meaning. In a world heavily reliant on visual imagery, bodily representation becomes one of the principal means of communication. This parallels strongly with sun-lust or recreational tourist experiences that centre on the body as a site of pleasure and performance. According to Markwick, coastal tourist imagery is 'focused on the body', with 'images of the body typically embedded within coastal scenes depicting leisure pursuits associated with the sun and fun of mass tourist activities' (2001, p. 424). Eden, Zeus and Greek Gold (which also double as spaces of liminal experience) are examples of a destinations focused on the body typical of coastal resorts. More so than Apollo, which evokes the picturesque and the private, the other spaces are set up for pleasure, particularly public pleasure. They are spaces 'about the body-as-seen, displaying, performing and seducing visitors with skill, charm, strength, sexuality and so on' (Urry 2002, p. 156). Pleasure is a display, with the environments set up for sexualised performance. The body in tourist spaces is experienced differently from in the home environment; the body is central to experience rather than being passive in the home and work environment. Bodily experience, specifically hedonism, is often acted out through club and beach environments where the body is the centre of display and the performance of one's body is the basis for interaction. In a study on British youths' tourist experiences of sun destinations, Diken and Laustsen (2004, p. 101) contend that 'the "exceptional" life of the tourist in Ibiza or Faliraki is not simply external to civilization; rather, the tourist now occupies a threshold in which life and law, outside and inside, civilization and state of nature become indistinct'. The appeal of this 'exceptional life' is a permission to do as one pleases.

According to Wearing, Stevenson and Young (2010, p. 66), 'there exists a subversive hegemonic bias towards masculine tourist experiences' where pleasure is defined as a masculine outcome:

> Studies of leisure and tourism experiences of both males and females indicate that sex-specific activities contribute to societal definitions of male and female identity, thus in many ways producing a conformity to gender stereotypes and acting as a restrictive rather than liberating influence on identity (Aitchison 2001, 2005; Samdahl 1992). Richter (1995) found that the linking of tourism with sex is rampant in marketing; for instance, souvenirs often promote women as sex objects, and destination attractions remain male-dominated preserves. (2010, p. 67)

The male gaze operates in different ways in the four spaces considered. On Apollo, the male gaze is normative: avatars appear gendered, often hyper-gendered, and are often found in heterosexual couplings. Many spaces on the island, such as the chapel, secluded areas of the island or Salsa y Boleros (shown below) are used as spaces for romantic liaisons of heterosexual coupled avatars. Similarly Eden, although a female-only space, is regulated by the male gaze even in its overt absence. As discussed in Chapter 5 (discussing the female body on Eden), the hyper-gendered form of many of the female avatars and the possibility of infiltration (by males in drag), in conjunction with the heteronormative design and structure of the space in reference to commercial sexualisation, signal the covert presence of male desire. This desire is both explicit and celebrated on the site of Zeus, where the gaze changes to include the male body as the object of pleasure, but is used as a means of opposition on Greek Gold, whereby female gendered and sexual expression can be seen as liberating only in opposition to, and through the absence of, the male gaze (Figure 2.1).

With the exception of Apollo, which is offered as an all-inclusive space to visit, the spaces analysed are specifically designated queer spaces, meaning they are specific sites for alternative sexualities. They are attractive for reasons similar to that of CL queer spaces in that they offer safe seclusion for sexual expression outside of normative culture. However, they are somewhat different to urban queer zones and gay

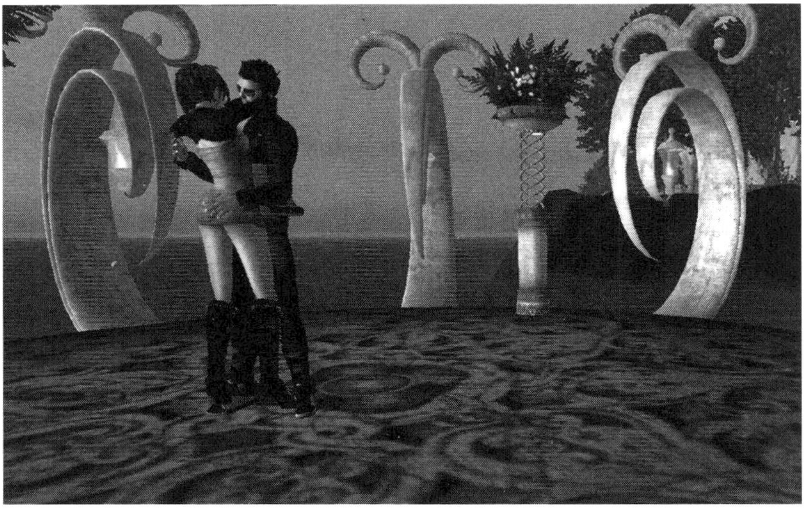

Figure 2.1 Salsa y Boleros, Apollo
Source: With permission from Linden Lab and the residents of Second Life.

ghettos in that they offer a tourist experience: travel to another place removed from the local where one can meet individuals from all over the world in a paradise setting, similar in structure to CL gay travel destinations (beaches, bars and clubs, and secluded zones for sex). As noted by Waitt and Markwell (2006, p. 117), the nightclub environment is persistent in queer travel discourse due to the sanctuary offered by the secluded environment:

> On the mainland the nightclub offers the gay traveller a controlled safe(r) environment, free from potential homophobia on the other side of the boundaries drawn by its doors. Inside, gay tourists are assured a familiar space where they can be themselves. The nightclub provides a place for the enactment of global gay identities through the presence of same-sex attracted people, bars, dance floor, music, gay porn video booths, and back rooms for sex.

It is unsurprising then that gay spaces in SL often have dance clubs as a primary part of the experience. Clubs not only provide a confined space for avatars to mingle and view each other closely, but also signify bodily performance, hedonism and, in particular, casual sex. The link between dance clubs and sex is often assumed, particularly in the gay community. This arose out of necessity in the 1950's when individuals had to be far more discreet about their sexuality and so sought appropriate meeting places. It has often been replicated since in mass cultural representations of gay life. Outside of the club, the beaches, gardens and other zones follow the discursive structure of gay travel media, and can be easily identified as queer. Uses of gay signifiers abound, from rainbow flags, to half-naked bodies, to advertisements for sexual services and products. This subversive sexual discourse, however, continues to operate 'within the structures of white heteropatriarchy rather than as a critique or a challenge' (p. 124) and continues to reinforce performance and interaction regulated through and by the male gaze. Zeus, for instance, with its reinforcement of commercialisation and its focus on the muscular and assertive male body, affirms the heteropatriarchy of male dominance and control, signalling a normative representation of form within a matrix of expression that mirrors heterosexual conduct and power.

Unlike gay male tourism, lesbian tourism has been minimally reported and researched, 'with very little academic literature exist[ing] about gay women and lesbians in terms of travel and tourism' (Waitt & Markwell 2006, p. 33). According to Puar, the gay female statistic is made invisible by the overarching categorisation of 'gay' that groups male and female homosexuals together, often disguising the disproportionate number of gay male to gay female tourists (2003, p. 938). Waitt and Markwell (2006, p. 33) speculate that this may be due to the following factors:

> it is likely that lesbians shoulder the burden of raising children more often than do gay men; that their income levels are on average lower than gay men's; and that they also are more susceptible to subtle and not-so-subtle forms of discrimination stemming from both sexism and homophobia than are gay men.

If these factors, speculated upon by Waitt and Markwell, are indeed valid, then the virtual platform may be a way to circumvent these barriers to lesbian tourist experience, given that the limiting factors stated include those of a geographical and economic nature. So, if it is not possible for a gay woman to escape the confines of her domestic/work space due to either domestic or financial impasses, or if she is too anxious about travel due to a fear of discrimination, then the virtual platform may provide a means of touring gay/queer/lesbian spaces outside of those in immediate geographical reach. Spaces such as Eden and Greek Gold, by only allowing female avatars, can be very welcome environments for women who have been harassed, or are fearful of harassment, due to either their gender or sexuality. Although there is the possibility for men to pose as female avatars in order to access the spaces, moderators are able to remove any avatar acting in an unsuitable manner. However, it is still possible for men to access the site if they behave within the guidelines imposed by the site, making it impossible to ensure that all avatars are being controlled by CL females.

Given the gendered nature of the environments, it is interesting to analyse the role of the body as presented on Eden within the parameters of tourist discourse. The Eden environment has been constructed for the pleasures of the body, in line with the bodily emphasis seen in coastal resorts such as the Mediterranean holiday island: there are beaches to sunbathe on and passively display the body; there are shops to purchase items to adorn and present the body; there is a nightclub and dance-floor to actively engage in bodily interaction and the active gaze; and there are lounges for more intimate forms of bodily interaction and activity. According to Markwick, 'Such imagery draws heavily on the popularly supposed oppositions of work and leisure' where the binary is 'played out in the imagery focused on relatively passive forms of leisure such as sunbathing' (2001, p. 424). The replication of the work/leisure binary in a virtual space is consistent with the desire to immerse the self in a space symbolic of pleasure; even if an individual is stuck in their work/home environment, it is possible to 'get away from it all' through the computer. In Eden however, what is notable for a female-only and lesbian environment is the replication of heteronormative bodily symbolism that encourage both passive and active forms of looking and being looked at. Here, the tourist gaze becomes centred on bodily performance, akin to that in heterosexual and gay male spaces such as Zeus. Whilst it may be possible that individuals hyper-gender their avatars in Eden to ensure that they are seen as female in a female-only environment, it is more probable that such appearances are due to fantasy and desire replicating mass consumption practices; bodily performance is of the desirable fantasy self which is open to eroticisation and the potential for sexual exchange. The self in an environment such as Eden is underpinned by its sexual use-value, with visual signification as the currency of this exchange. As discussed in Chapter 5 (discussing the female body on Eden), however, this hyper-gendered form of femininity is problematic in terms of the potential of infiltration by 'males in drag' as well as diluting lesbian performativity to more acceptable and normative performances. Whilst Eden can be viewed as an extension of the heterosexual matrix, Greek Gold is unique as a gendered and sexualised space constructed for the female gaze. Unlike the normatively presented and hyper-gendered representations of females (and males) on the spaces of Apollo, Eden and Zeus, the representations on Greek Gold are far more diverse with a fluidity of gender expression attached to the female subject. This diversity and opposition to the established male=gaze of space links the region most readily to ideas of Lesbos and Sparta and to notions of returning to a nostalgic, mythical past.

It is important, even in the increasing acceptability of alternative sexualities globally, for there to be spaces designated as queer. Alongside the marketing and commercial aspects of tailoring to a specific market segment, such spaces are enriching for gay individuals by offering identity, safety and community. This is discussed by Pritchard, Morgan and Sedgley (2007, p. 276) who contend that gay and lesbian spaces are empowering for the homosexual community, but note that they are becoming 'subject to heterosexual attack, either via abuse or encroachment leading to the dilution and erosion of gay and lesbian spaces and identities'. The potential for 'dilution' of queer space by heterosexual interlopers can happen covertly, through the CL male embodiment of a female avatar on Eden or Greek Gold, or less covertly through heterosexual presence on Eden, Greek Gold or Zeus. According to Casey, there is a recent trend for heterosexuals to visit queer spaces, suggesting a greater recognition of queer lives, but also illustrating the fragility of such spaces in terms of identity and solidarity (2004, p. 448):

> Perhaps one of the most recent developments that not only reflects the commercial success, but also the continued fluidity *and* fragility of spaces within the 'gay village', is their increasing popularity among heterosexual visitors and tourists, acting as attractive spaces to enter and consume.

Such fluidity may not reflect an acceptance of alternative sexualities, however, as it may be that heterosexuals are entering queer environments in order to view the performance of the 'other': in the sense of a theatrical performance of the exotic. As detailed in Chapter 5 (a discussion of the female-only space of Eden), such spaces are becoming cosmopolitan tourist spectacles for heterosexuals, particularly heterosexual females who are able to use the spaces to get away from the male gaze whilst consuming queer (particularly gay male) culture as a fashionable and exotic experience. Such dilution can be problematic for gay spaces in general, but poses specific problems for queer female space. Heterosexual touring of Eden, for instance, can take on a more inauspicious characteristic with primary reasons for visiting the space being either surveillance of the 'other' or a desire to participate in a hetero-patriarchal fantasy of lesbian sex, thus undermining the site's relevance as a female-only lesbian venue. The evocation of the gaze, and the centrality of the sexualised body within a lesbian environment, is a cultural shift away from the conservatism often associated with gay female culture and its notions of deeply emotional and monogamous relationships. Whether the hetero-patriarchal structuring of space and the elicitation of the sexualised gaze, typified by Eden, is in fact empowering for the lesbian community is questionable. On the one hand, it can be empowering for women to be consumers of the sexualised gaze; on the other, it may be that such a gaze remains chiefly in the power of the heterosexual patriarch.

Conclusion

This chapter has described the SL environment of Apollo, Zeus, Eden, and Greek Gold as tourist spaces. This is due to their conception as escapist, being either liminal, liminoid or both. Principally, the action within these spaces relies on the pleasures of looking (the gaze) as well as the presentation of a desired body. Practices of looking are premised on the environment and landscape as symbolic, evoking the ancient Mediterranean and its cultures. It is through this imagining that the liminal can be reached as in the corporeal journey of transcendence. That transcendence may be possible requires a re-thinking of our understanding of authenticity and what it means in regard to the virtual environment – the natural word and the sublime are analysed in relation to this phenomenon in the next chapter. On the other hand, the liminoid is easier to behold within the SL environment given that avatars commonly display desired (Western) attributes of the body. It is this focus on the body, and its pleasures, that is most commonly associated with SL and the four regions studied, in particular Eden and Zeus. Anonymity and desired virtual embodiment work together to produce a space that is a perfect periphery of liminoid behaviour. The sun, sex, sand and sea of CL are reproduced through the screen with similar results that reaffirm gender and sexual politics. How gender and sexuality are both constructed and consumed is discussed in detail in the next four chapters which explore each region in depth.

3
Intersections in Space, Nature and Mythology

Introduction

The Lost Gardens of Apollo (Apollo) is an island region that is visually spectacular, featuring mountains, gardens, ornamental buildings and stunning vistas. It is a space that readily evokes Western notions of paradise. By combining visual splendour and fantasy there is a reinscription of power predicated on the Western corporeal tradition. Space and nature are framed by historical and cultural conceptions of masculinity predicated upon patriarchal assumptions. This chapter is an analysis of the island of Apollo through the major categories of space, nature and mythology that reveal a place that is highly masculine and patriarchal in representation. Apollo's spatial practices reveal a similarity with the corporeal world, even in the knowledge and practice of different forms of embodiment and movement. Space within the virtual is obviously produced, arguably making the power intrinsic to the virtual environment more visible – there is little scope for falling back on natural and essentialist assumptions of the world. Similarly, the natural landscape as represented on Apollo (its mountains, gardens, flora and fauna) is evoked in much the same way as natural features in the CL environment, and so elicits an emotional and mental affect much as the Kantian notion of sublime works in corporeal nature. Exceeding the initial application of Kant's sublime, the virtual landscape of Apollo traverses several categorisations of the sublime in being dynamic, technological and digital. Mythology intersects with concepts of nature and space to generate meaning on the island. The overt mythology of Apollo is grounded on the ancient Grecian ideal of masculinity, male beauty and privilege, and the potential for male-centred and homophilic (same-sex) sexuality. This potential, however, produces a tension on Apollo: that of an open and proud homosexual identity and one that is hidden, codified by the consumerist practices of the postmodern Western world. Arguably, it is the latter condition that remains the more powerful and ultimately more visible in its invisibility, subsuming a homosocial and homoerotic otherness within a hyper-masculinised and heterosexually assumed identity. The modality of visual culture here is investigated through the representational aspects of identity, analysing the body as object and as potentially subversive.

Space

According to spatial theorists working in the fields of cultural theory (such as Lefebvre, Foucault, Soja and Massey), space is produced; it is a social construction that both acts and affects subjects whilst being acted upon in return. Ideas about space as produced are easily applicable and understandable in the context of virtual worlds such as SL, which present consciously produced environments. Whilst it may require a certain active consciousness to identify the power relations of supposedly objective space in the corporeal world, it is difficult in the virtual to sustain such a false consciousness. Space, as we perceive it, is a creation bounded by social, political, historical and cultural influence. According to Prigge, the analysis of space in reference to symbolism and representation can illuminate the significance of space production and reception as politically contextual, destabilising the idea of a naturalised space (2008, p. 48):

> If social power is symbolized in the appropriation of space, the significance of such spatialization is revealed only through an analysis of these relations as relations of meaning ... How the imaginary and the symbolic character of the spatial is historically inscribed in social contexts remains concealed, therefore, if it is not theoretically reconstructed as a real object of discursive and symbolic practice and therewith recognized as a specific form of constituting the social. Such an analysis must clarify the daily dominance of spatial concepts as the discursive hegemony of spatial modes of speech.

In the constructed space of Apollo, space is obviously symbolic. Meaning on the island is produced through relational signifiers that draw upon the social practices of corporeal existence. Furthermore, these practices are specifically laden with power relations, so that power is constituted both through and within spatial practices that mirror CL. It is these reflective spatial modes, ones it could be said that enact 'real' experiences, which reinforce the discursive power structures in-world. As a virtual space, Apollo is defined through the corporeal and is representative of the cultural paradigms from which it promises an escape. The rules of the 'real', of the everyday, are inscribed upon the virtual, in a fantasy world that is as much anchored in discursive practice as the daily life of corporeal existence.

There are explicit ways in which Apollo can be used, in the sense that it is experienced, that are aligned with the intentions of the creator and administrators of the island space. In addition, there are usages that are implicit, or governed by the conventions of space experienced in the corporeal realm. For instance, there are many pathways that lead an avatar around the island for the purposes of exploration or sightseeing. In addition, there are structures and features of the island that lend themselves to spatial practice in a similar way to which corporeal spaces are experienced. On teleporting to the island, one's avatar is positioned at Hyacinth Valley, from where a number of garden pathways lead out towards various landmarks on the island. To take these paths as a basis for exploration is to utilise the space as it has been inscribed, as dictated and governed by the normative practices of movement. These practices are well established in the corporeal, in what de Certeau describes as a Foucauldian apparatus of power: 'The pathway could be inscribed as a consequence, but also as the reciprocal, of Foucault's analysis of the structures of power', so that 'spatial practices in fact secretly structure the determining conditions of social life' (1984, p. 96). On Apollo, spatial practice, and thus social life, is governed by corporeal discourse. How an avatar uses the space is closely aligned to the real, with the exceptions being the forms of movement that depart from 'realistic' spatial appropriation, such as teleporting or flying.

As a result, action and interaction in the virtual realm are bounded by the very restraints it promises an escape from. As much as the utopian view promises such escape, it is not possible to view virtual space as a separate space from the corporeal; they act symbiotically on each other. This is not to say that virtual space cannot be emancipatory or exist at the margins of possibility, for it is this very marginality that presents the virtual as so attractive to so many. Furthermore, it is the space's lack of concreteness and exactness that allows for reflection and different modes of engagement with the world and the self, and such re-imaginings can be considered as having social, cultural and political significance. As explained by Prigge, virtual spaces can provide a departure point for the reification of the self (2008, p. 54):

> Ideological confrontations in everyday life seek to increase the[ir] respective room to maneuver and the[ir] articulation of interests by shifting the frontiers between dominant and dominated spaces. One possibility for increasing this room to maneuver lies in aesthetic spatial practices, where the power of images can reach beyond set frontiers towards utopias free of dominance. This allows people to imagine possibilities banished from the realm of ideology.

The possibilities of new and different aesthetic spaces allow for a reinvestment in the inscription of the self in relation to the aesthetic environment. One does not simply have to accept the status quo, and/or the prevailing attitudes and norms, of the pre-existing corporeal world. Consciousness necessarily shifts to encompass new ways of being, as ontological and phenomenological tropes of the self are elicited through virtual embodiment and experience outside of those bounded by the corporeal. It is not possible, however, for the self to be fully removed from the structures of the corporeal: 'one is not therefore in the realm of truth and outside ideology, for there are no true and non-ideological forms of knowledge' (Prigge 2008, p. 54). Thus, virtual spaces may be at the margins of ideology, where the imaginary and corporeal intersect in symbiotic exchange. As the example of Apollo shows, the corporeal is consistently signified in the representations of landscapes and nature, reinscribing notions of power onto the environment. The imaginary, both as protagonist and antagonist to existing power relations, often draws upon ideas of myth and mythology to iterate meaning.

The natural world and the sublime

To describe a virtual landscape as 'natural' seems entirely contradictory: the natural world is considered to be that which we can touch and feel: not made by us, but detached. However, as described by Rod Giblett, there are classifications of nature that encompass a spectrum of conceived nature, and which include the hyper-real of image and simulation. Describing the degrees that separate the natural object from the indigenous environment, he classifies the hypermodern mediated environment as 'fourth nature', where first nature is indigenous and the proceeding classifications are shifts away from this state (2011, p. 23). Apollo is certainly not natural in terms of the first nature classification, and most readily fits into the fourth nature definition, or '"the new nature" of the "feral future"' (Giblett 2011, p. 23) which promises real-time access to everything. The immediacy and acceleration of landscapes, of fourth nature, have been part of our globalised world where communications infrastructure has allowed for the transmission of exotic and sublime images into the recesses of the everyday. Nature is no longer something that is out there – mass accessibility has come about through the creation of desire (nature as a mass consumable) and through mobility (not only through the physical sense of global travel but also in class mobility and the associated increase in leisure options). Moreover, nature is now something that can be experienced almost entirely through mediated activities. Although principally reliant on visual stimuli, the virtual worlds of the internet now dominate the ideas of the natural environment, including nature as spectacle and sublime.

Fascination with landscapes can be described through Kant's notion of the dynamic sublime as well as the idea of the extraordinary as contrasted with the everyday. Kant's dynamic sublime details the awe that we experience as humans when confronted with a nature that is so large and all-encompassing as to render our own existence insignificant. The feelings evoked by this confrontation are fear and terror in reaction to our inability to dominate the landscape, but also admiration and reverence in response to the recognition of something so extraordinary – the sublime reinforces our cognitive capacities, thereby reasserting our position as humans. The virtual landscape obscures these ideas of the natural sublime as our relationship with fear and terror is subdued by our superficial mastery of the environment. Our feelings towards the natural landscape (first nature) are replaced with feelings of the technological sublime. David Nye describes this shift as follows (1994, p. 282):

> There is an American penchant for thinking of the subject as a consciousness that can stand apart from the world and project its will upon it. In this mode of thought, the subject elides Kantian transcendental reason with technological reason and sees new structures and inventions as continuations of nature. Those operating within this logic embrace the reconstruction of the life-world by machinery, experience the dislocations and perceptual disorientations caused by this reconstruction in terms of awe and wonder, and, in their excitement, feel insulated from immediate danger. New technologies become self-justifying parts of a national destiny, just as the natural sublime once undergirded the rhetoric of manifest destiny. Fundamental changes in the landscape paradoxically seem part of an inevitable process in harmony with nature.

Although writing from an American perspective, Nye's ideas are readily applicable to the idea of technological mastery of the entire Western world, where technology is seen as instrumental to our collective human destiny. Utopian ideals spring forth from our mastery of the natural environment, replaced by technologies and associated feelings of awe, reverence and hope for the future. Fear of the natural environment is replaced with fears of human-made designs and confrontations. No longer do we fear insignificance in relation to the natural landscape; we now fear alienation and irrelevance in our mediated technological world (Figure 3.1).

The technological sublime runs parallel to the natural sublime in Apollo. Whilst the feelings of awe and reverence for the natural landscape are replaced by the feelings about technology, the landscape proclaims the extraordinariness of human vision and potential. Virtual landscapes have become part of the new, accelerated sublime of tourism and experience. According to Claudia Bell and John Lyall, the accelerated sublime is part of the postmodern age of increased vastness, speed and consumption, changing to encompass new objects and landscapes as they are discovered. In discussing landscapes, they argue that 'representations are not an accurate substitute' of the sublime in that the

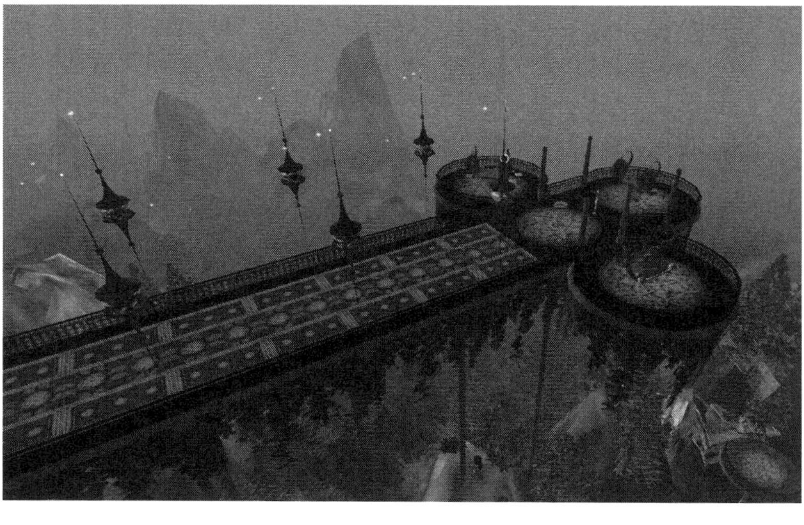

Figure 3.1 The Bridge to Nowhere, Apollo
Source: With permission from Linden Lab and the residents of Second Life.

representation loses scale and therefore awe (2002, p. 199). This is a contestable position, given that one's embodiment as an avatar on Apollo can elicit feelings of insignificance in reference to the constructed landscape. In Figure 3.1, the structure of The Bridge to Nowhere is far larger than any avatar, to the point that its scale does in fact produce an affective state. Similarly, the East-West Towers are so high in comparison to avatar dimensions that one must fly in order to view them. In a comparable fashion, Apollo Harbour is a place of such detail, scale and splendour that moving through the environment also produces affective states. Whilst these feelings may not be fear or terror in reference to the dynamic sublime, it is arguable that many individuals visit the island of Apollo for the beauty of the environment, and the associated feelings of veneration, much like those induced by first nature.

As pointed out by Vincent Mosco, the sublime and myth exist side by side, and in doing so propagate the hopes and fears of humanity. He describes cyberspace as 'the latest icon of the technological and electronic sublime, praised for its epochal and transcendent characteristics and demonized for the depth of the evil it can conjure' (2004, p. 24). The digital sublime advances the myth of cyberspace as utopian – a world devoid of the banality, restrictions and judgement of the corporeal. It also, however, simultaneously elicits fears of the unknown and the uncertain, specifically in relation to governance, control and data ownership. In a more general sense, there is the myth of the end of history that is associated with digital technology (Mosco 2004, p. 117):

> [O]ne cannot help but conclude that the rhetoric of technology, the technological sublime that David Nye so perceptively identifies, is powerful enough to create a widespread historical amnesia. One of the more useful ways to understand technological myths, including myths of cyberspace, is to excavate the tales that accompanied the rise of earlier 'history-ending' technologies.

With new technologies come the perpetuated myths that humanity can transcend our existence to reach a heavenly, or more perfect, space. What virtual environments more specifically promise, however, is a visual simulation of myth, so that one can visually immerse oneself in the paradise offered. It is this facet of the digital sublime that is so compelling for users.

There are many parallels between the ideas of heaven and the virtual worlds offered within cyberspace. One of the major attractions of Christianity, as discussed by Margaret Wertheim, was its openness to all, where 'people of all nations and skin colours can aspire to walk the streets of the Heavenly City' (1999, p. 23). So, too, is the acceptance offered within the virtual environment. The region of Apollo, for instance, explicitly states that it is a space 'for all creeds' and values tolerance and acceptance (Second Life places 2009). Here, the ideals of Heaven are reaffirmed in contrast to the 'host' culture from which it is derived. The digital idea of heaven is not solely drawn from the Christian concept, however, with notions of paradise evoking various mythologies. Apollo's heaven is most readily aligned to that of the ancient Greek (particularly Athenian) ideal. The aesthetic pronounced in bodily forms and the sublime is conjured through the technological mastery of the island features and the detail and vastness of the natural (virtually reproduced) environment. The technological and dynamic sublime are evoked on Apollo, increasing the awe and reverence evoked by the space, ultimately positioning paradise as spectacle. Attention to the body beautiful not only resonates with Grecian ideas of perfection, and the perfect body, but also reinscribes the ancient Grecian ideal onto the present (or, as described below, it could be that the reverse is true). What is ultimately apparent in the imagined paradise of Apollo, however, is the reproduction of the classical masculine ideal.

The ancient world and mythologies: Affirming masculinity

Masculinity in the Apollo region is chiefly inscribed through representations of ancient Greece and its surrounding mythology and conception in the popular imagination. The ideas of maleness and modern masculinity have historical roots in ancient cultures that privilege the male form as active, muscular and powerful as well as beautiful. To uphold gender divisions, and therefore reinforce the position of man, it was necessary for Athenian male citizens to continuously perform masculinity bounded by the state and its norms. According to Matthew Fox, scholars of the Athenian male suggest that he was not a free subject but was constrained by the structures of society that necessitated masculine performance: 'the citizen is bound by the conventions of his society both to exercise mastery over his social inferiors, and to devote the same effort to mastering himself' (1998, p. 7). In drawing upon the work of Winkler, Fox explains (1998, p. 7):

> The citizen elite consists of individuals dedicated to maintaining their reputation; to penetrating others, retaining identification with

> an integrated and inviolable phallus, and warding off any suggestion of being like a receptacle for the sexual or social desires of others.

Fox maintains that these accounts of history have a tendency to view ancient societies within the established framework of contemporary political discourses, especially in reference to sex as a key indicator of identity. This self-reflexive understanding of the ancient world through current perspectives is illustrated in the popular imagination: 'The Real confirms the partial quality of all discourse' (Fox 1998, p. 18). Our understanding of ancient cultures, then, is always, to varying degrees, mythological. That such myths are reproduced in the virtual environment is fitting in the consignment of truth to imagination.

So, what is of importance in the creation and appropriation of the spaces is the collective imagination (that of mass Western culture), in terms of Westernised patriarchal discourse. Conceptions of ancient culture within the virtual are represented strongly through the symbolism on Apollo, in the objects, buildings, structures and architecture of the island which often evoke a masculine form. The island's name connotes the mythological and the male which the other representations on the island signify. The ancient Greek discursive practices of male dominance, hero-worship and homosexual infatuation and Eros are played out in the myth of Apollo and Hyacinth (Hyacinthus) (Harris & Platzner 1998, p. 169):

> In the god's most celebrated affair, an outstandingly handsome boy named Hyacinthus is the object of Apollo's desire ... While throwing the discus together, a favourite pastime for Greek lovers, Hyacinthus, perhaps accidentally, steps in the way of Apollo's hurtling discus, which strikes him fatally in the head. (One version of the myth states that Zephyrus, the West Wind, also loved Hyacinthus and, when rejected in Apollo's favour, jealously diverts the discus's trajectory, causing it to strike the young man).

Mourning the death of Hyacinthus, Apollo named the Hyacinth flower after him. Within SL the Lost Gardens of Apollo are symbolic of the classical myth. Although Apollo had a number of female lovers, it is the love of Hyacinth that is foregrounded on the island, and is the most well known of Apollo's eros in contemporary Western literature. The island of Apollo honours the ideal of the beautiful young male as represented by Hyacinth. The valley that takes his name, Hyacinth Valley, is the entry point onto the island; avatars generally teleport to Apollo through the hubs located in the valley. That the island's name and its central region are named after the myth of two beautiful young male lovers leads to the appropriation of this central signifier to all other aspects of the island itself, signalling participation in the narrative and its associated positioning of the male as beautiful (Figure 3.2).

The male body in ancient Greece was understood as an object of beauty: 'the norm, against which those of boys, females, slaves and barbarians were all seen as deviations to a greater or lesser degree' (Stewart 1997, p. 11). The statues of Apollo epitomise the perfection of male beauty; they are muscular, controlled, and dominate the landscape. Graceful in their suspension, the male statues of Apollo (shown above) represent the physical ability and strength of the male subject, youthful and virile: in Greece, 'the sculpted naked body was a harmonious design that illustrated divinity and strength' (Brownmiller 1984, p. 23). Each sculpture faces the other and they almost touch at the torso, invoking homoeroticism, a practice thought prolific in ancient Greece. According to Whitley, '[o]f all Greek art forms, sculpture was held in the highest regard' (2001, p. 270), and statues 'were idealised representations of how gods or heroes ought to appear, using the language of realism

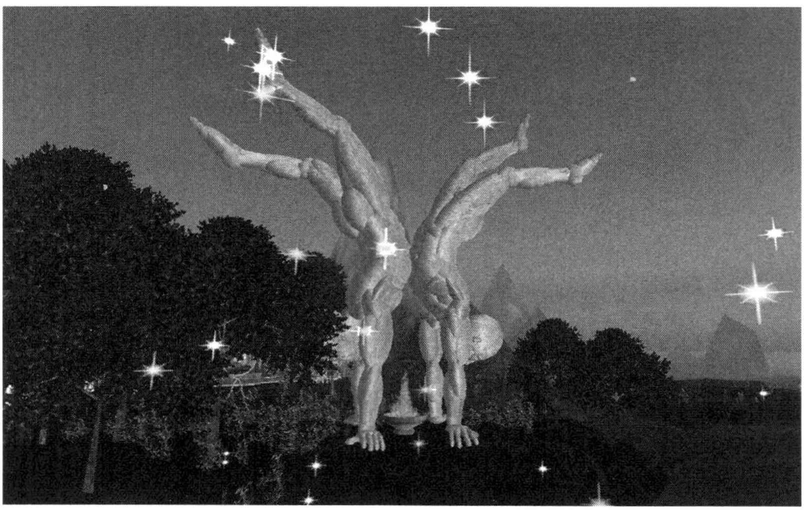

Figure 3.2 Male statues, Apollo
Source: With permission from Linden Lab and the residents of Second Life.

without in any sense being realistic' (Whitley 2001, p. 278). The Apollo statues evoke the ideal male form while reinforcing the contemporary normalisation of the male body as commodity and presenting examples of an ultimate standard of aspiration. Moreover, that the statues snub the principles of physics suggests a mastery over the constraints of science in the realm of fantasy. The masculine ego can become uninhibited in virtual space that is unconfined by the laws of physics or science. The ego can acutely display a mastery over the environment as it projects itself into the future where it can become even more masterful in its control. In contemplating art and science as they correlate to the male ego, Antony Easthope recounts the psychoanalytical view that artistic expression relates to sexual drive as 'a form of sublimated desire'. Moreover, 'scientific projects, the attempts to master nature through knowledge, are explained in terms of narcissism and the ego' (Easthope 1992, p. 39). Thus, the Apollo statues evoke both 'drives' of masculine ego; they are structures that exhibit the ideal of masculine perfection and masculine homoeroticism as well as being symbols of male control, mastery and power.

Broader contextualising of masculine fantasy and myth evokes ideas of the natural self, isolation and separateness where the stagnancy of civilisation and CL can be transcended. According to Gregory Woods, such exploration provides 'fantasies of pre-industrial peacefulness and prelapsarian sexuality'. Islands such as Apollo allow the body to 'resume its "natural" condition' (1995, p. 126). Woods identifies a key narrative of the island fantasy that closely aligns thematically with Apollo (1995, p. 126):

> isolated males form a relationship with the landscape and the elements, then relate homosocially and homoerotically to each other, in febrile renegotiations of their masculinity, before returning to white heterosexual civilisation.

In light of the fact that Apollo is a PG-rated island, it can be assumed that avatars do not visit Apollo for overt sexual purposes (however, such visitations may well be sexually covert in negotiating identity); so, given the homoerotic nature of the island, Woods's notion can be made applicable to a masculinised relationship with Apollo: users may log on to SL for escapism and during this time renegotiate their identities within the island narrative of a pre-industrialised, pre-civilised landscape. Such renegotiations may well be covertly sexual and may be projected onto the landscape (Woods 1995, p. 129). On Apollo, this projection takes form in the 'natural' landscapes that are occupied and the masculinised objects that are observed; the feminine is controlled and appropriated whilst the masculine is uninhibited and displayed.

Sexuality and mythology: Interpreting the male subject

In addition to embodying an idealised form of masculinity, the experience of Apollo as a mythological site, and its construction as a space of alternate expression, allow for ideas of sex and sexual experience outside the regulation of corporeal society, drawing upon the imagined and actual practices of ancient behavioural norms. Homosexuality has been attributed to the classical world since the 1970s and the rediscovery of same-sex eros in classical texts. Ancient Athenian practices of what we would now call homosexuality were predicated on the homosociality of citizen interactions (citizens being a certain class of adult males) and the ideal of the masculinised form as the most desirable. According to Stewart (1997, p. 10):

> This kind of institutionalised homosociality conforms exactly to the classic definition of patriarchy: 'Relations between men, which have a material base, and which, through hierarchy, establish or create interdependence and solidarity among men that enable them to dominate women' – and in the Greek case, we may add, everyone else as well.

By 'everyone else', Stewart means everyone who did not have full citizen rights in ancient Greece, and therefore were not an accepted part of the patriarchy. The citizen was an ideal male who possessed the finest qualities of the masculine. Moreover, the state acted as an institution 'that treated adolescent boys as minors along with females and slaves, both of which were already seen as natural receptacles for the citizen's roving desire' (1997, p. 10). Male homophilic conduct was a normalised part of ancient Greek life, 'since the Greeks saw nothing surprising in the co-existence of desire for boys and desire for girls in the same person' (Dover 2002, p. 26). Dover states that it was commonplace for older citizens to pursue beautiful young boys as much, if not more so, than women. Such conduct was regulated by an established class system that had heavily codified behavioural standards. A citizen:

> could seek among his juniors a partner of citizen status, who could certainly not be forced and who might be totally resistant to even the

> most disguised kind of purchase. If he was to succeed in seducing this boy (or if later, as a mature man, he was to seduce a youth), he could do so only by earning hero-worship. (2002, p. 26)

Homophilic, particularly pederastal, conduct was assumed under the guise of a mentor-like bond between males, with the older male earning the youth's or boy's infatuation through status and accomplishment. Often, such eros was considered of greater social worth than heterosexual relations, given the superiority of the male in ancient Greek society. There are, however, debates in relation to the full acceptance of homosexual and pederastal conduct in classical societies, such as Thomas Hubbard's (2009) study examining the debates about 'natural' sexual conduct pre-dating Plato. The accuracy of the contemporary appropriation of ancient cultures in virtual worlds is ultimately inconsequential, as history is imagined and mythologised.

Although there were no instances of pederastal conduct detected on Apollo (or on Zeus), what the space does epitomise is a fascination with youth and beauty as illustrated in the appearance of the male avatars and in the imagery found within the spaces. Moreover, there is an attraction to the de-normalisation of sex and sexual practices as they relate to the fields of normative sexual and relational conduct in Western societies. As discussed further in the chapter on Zeus (Chapter 4), there are fears that unabated sexual desire, as culturally ascribed to men, would lead to the erosion of civilised society that is based on the family unit. Interestingly, there were arguments in ancient Athens entirely to the contrary. In outlining the debates regarding homophilic and pederastal conduct, Thomas Hubbard illustrates the divisive quality of male sexuality in ancient societies. Although tolerated, and sometimes celebrated, homosexual and pedesteral conduct was seen as unnatural by some critics of homosexuality in ancient Athens. Supporters, however, saw such conduct as a cultural development that surpassed nature by illustrating human control and development, 'presenting the stages of human progress as a continuous ascent from vulnerability to the elements to satisfaction of basic needs to more advanced forms of political organization and technological sophistication' (Hubbard 2009, p. 255). In looking at the moral panics in ancient and contemporary culture, paradoxes regarding the notions of nature and masculinity are evident. Masculinity and sex in the classical world were seen as cultured, as something to be developed in opposition to nature: 'Culture emerges precisely out of the male principle's fear of and struggle against the formlessness of female Nature, either by attempting to objectify female Nature into an ordered artistic recreation ('Beauty') or retreating from the feminine altogether (homosexuality)' (Hubbard 2009, p. 256). However, contemporary moral panics regarding masculinity and sexuality fear an essentialised nature of masculinity, particularly in reference to sex. To be cultured is to be in control of one's desires; to allow one's sexual nature free reign is to court danger and risk destabilising the family and relational structures of society.

Contemporary masculinities: Contextualising the male subject

A tension exists between Western standards of conduct required by the masculine subject and the subversive signification of the homoerotic, and potentially homosexual. Apollo, despite appeals to the contrary, is much more a place of patriarchal fantasy than an environment of anything-goes gender and sexuality. This is due in part to its PG rating 'where two or more of any orientation can non-sexually cuddle and enjoy the surroundings' (Zander as cited in James Au, 2006). Although there are many examples of homoerotic signification throughout the island, much of it can be read through the lens of the contemporary sexualisation of male beauty. It can be argued that the objectification and sexualisation of the male body on Apollo invites the female gaze and is thus empowering for women in reference to men. However, due to the extent to which the masculine form is prized on Apollo, there is also a sense of a competition of beauty and of being noticed that embeds the gaze as male. On Apollo, as in ancient Greece and gay male culture, it is not necessary for the female to display beauty, to be objectified, or even be present at all. The significance of beauty is of and through the representation of a masculinised ideal that moves between the homosexual and heterosexual.

Representations of the homoerotic on Apollo range from displays of art through to the performance of avatars and their symbolic depiction. Male avatars often appeal to the Westernised norm of attractiveness and are frequently dressed conservatively. Some male avatars, however, exude more sexuality in their appearance, with exposed muscles and what could be considered as more fashionable, and perhaps flamboyant, attire. In encouraging the gaze they epitomise the contradictory construction of homoeroticism: they convey the ideal of the masculine form but in doing so can feminise themselves as objects of desire. Although such eroticism and desire may be grounded in heterosexuality, the display of overt masculinity is a performance of the self – inasmuch as it is comparable to the performance of others and so correspondingly encourages the subject to gaze on others as he is gazed upon. According to Rohlinger, male attractiveness has become commonplace in society due to cultural and economic trends shifting to a greater commercialisation of the body (2002, p. 62):

> In recent years, sexualized images of men, or the 'erotic male,' have proliferated in men's magazines. In these images, the erotic male represents a physical and sexual ideal, whereby an attractive, muscular man is placed on display. Such imagery is undoubtedly in part a response to the economic trends over the last 50 years, but it is also a product of cultural changes in American society.

In her analysis of magazine images, Rohlinger notes that there has been a significant movement towards male representation of 'unknown sexuality'. This ambiguity points to the commercialisation and commoditisation of the masculine in Western society, and not only signifies a growing acceptance of the homoerotic form, but also the objectification of the male heterosexual body.

However, displays of 'unknown sexuality' as described by Rohlinger also illustrate a continuing unease about overt homosexuality within society, particularly as characterised by gender subversion. On Apollo, as in Rohlinger's case study on magazines, sexual and gender expression adhere to the cultural standard, so that the appeal to the homosexual gaze becomes coded, where the subtext lies in the tension of looking and being looked at. Sexual ambiguity evokes the unease that comes from the transgression of gender boundaries; there is an unknown quality in the looking-ness, in both application and reception, as the gaze moves undeciphered through the spectrum of homosexual, homoerotic, homosocial and heterosexual connotations. The movement between heterosexual appreciation and homosexual desire of the male body not only evokes a tension of sexuality but also the anxiety about the masculine performance of the commoditised body. As a reflection of corporeal society, Apollo highlights the homosexual/homosocial continuum, whereby the homosexual is an accepted facet of culture in its codified and more covert representations. The hyper-masculine may also become commoditised, like the female body, but it is not feminised insofar as it does not take on the same level of objectification and subjugation, and therefore is able to maintain its status as an expression of power. Homosexuality, then, is both present and absent on Apollo, and often becomes more identifiable as homosocial masculine display. According to Sedgwick's understanding of the homosocial, 'masculinity is seen quite literally to depend on both the permanent presence and indeed absence of homosexuality' (as cited in Edwards 2006, p. 95). Masculine expression is defined through the tension between the ever present possibility of homosexuality and its repudiation. This acknowledgement and dismissal of homosexuality positions gay males, and even males of atypical masculinity, in a tenuous position within society: 'gay men are neither more nor less 'masculine' or misogynist than straight men, but located in an awkward, and perhaps even dialectical, relation to gender both psychologically and socially' (Edwards 2006, p. 96).

Femininity and the patriarchal assumption of space

The gendered assumptions of corporeal space are replicated in the visual and functional characteristics of Apollo. Representing two distinct, but overlapping, spheres of nature and technology, the island reinscribes notions of power within spatiality and representation. First, there is the construction of nature, albeit fourth nature, which replicates ideas of paradise through the screen: manicured gardens, lush greenery and audacious animals. This is nature as female, as the binary opposition to masculinised culture. The idea of nature as eternal, wholesome and beautiful is asserted through the various landscapes of Apollo. Even the sublime elements of nature, such as the central mountains, are seen as non-threatening and ultimately consumable. This is nature's place in view of the masculine mastery of computer generation; nature has been tamed, domesticated. Furthermore, technology and digitisation are the vestiges of maleness, enacting the 'male gaze of domination' (Nye 1994, p. 283). David Nye remarks on the ultimate feminisation of technology as it becomes part of the everyday, and as the sublime becomes commonplace, stating that 'Each form of the technological sublime became a 'natural' part of the world and ceased to amaze, though the capacity and desire for amazement persisted' (1994, p. 284). The sublime remains the privilege of the masculine gaze; the discoveries, the awe, the reverence, are reserved for those who dominate and master their environment in the continually evolving (arguably devolving) enterprise of progress. Consuming and producing space, then, are irrevocably masculine notions. The feminine, therefore, is the object of the gaze and of consumption.

Although the symbolism on Apollo is significantly masculine, there are some elements of feminine expression, albeit within the established confines of spatial regulation. The major locale for this expression is at the centre of the island, within Hyacinth Valley, which is surrounded by mountains. Only a few passages lead out to the coastal regions of Apollo, dissecting the sheer cliffs that surround the valley. Much of the coastline is typified by cliffs and rock-faces, or 'man'-made structures, such as the male statues or Apollo Harbour. These elements combine to present a mastery of nature, with the most masculine features at the island's edge, encapsulating a more feminised centre. It is as if the feminised beauty of the island, including the allusions to the myth of Hyacinthus and the eros shared between him and the god Apollo, necessitate protection and containment from the more masculine extremities of the island. The gardens within Hyacinth Valley and elsewhere are other examples of this containment. Although they may typify a feminine aesthetic, they are equally controlled and ordered through the spatial configuration of the island; they can be looked upon as elements of beauty without encroaching on the principles of orderly space or movement.

As part of the ancient Greek symbolism there is a particular representation of a goddess in the centre of Hyacinth Valley, where stands a welcome statue of a white goddess rising, or perhaps escaping, from a distorted figure of blackness. In ancient Greece, the wings of a god or goddess represent freedom and a magic or mastery over the mortal world: 'Prized Greek qualities of swiftness and competition are symbolized by winged creatures' (Hill 2006). It could be argued that the statue symbolises a freedom of spirit in the virtual world of SL, and in particular Apollo, where individuals can escape the darkness of the real world and transgress the boundaries of society. Or it could be argued that the goddess represents the tensions of gender and power in historical narrative. In the mythology of ancient Greece, the goddess represented the ideal female form, but was often idealised as very different from the standard woman (King 1993):

> For the Greeks woman is a necessary evil ... an evil because she is undisciplined and licentious, lacking the self-control of which men are capable, yet necessary to society as constructed by men, in order to reproduce it.

The goddess was a different type of woman, more aligned to the male/patriarchal area of the heavens than to the inferior female domain of Earthly servitude. They were 'immortal, ageless and powerful' (Lefkowitz 2002, p. 325). The goddess depicted on Apollo could be Eos, 'the goddess of the morning red, who brings up the light of day from the east' (Eos 2008), rising out of a figure of darkness that could possibly be Nyx, the goddess of night who lives 'in a palace constructed of a dark cave where the sun never shines' (Hill 2006). Eos, according to mythology, had a desire for beautiful male youths. Whilst the fascination with male beauty within contemporary society is generally accepted, as is an expressive female sexuality within accepted norms, in ancient Greece an expressive and dominant female sexuality was seen as a real threat to patriarchal society. Andrew Stewart explains that the story of the female goddess Eos and her desire for abducting youths could have been meant as a warning in Greek society: 'not only do these pictures hint at the evils of female dominance (*gynaikoarateia*) and easy capitulation to desire, but nervously evoke their appalling consequences: female control of the phallus' (as cited in Lefkowitz 2002, p. 327).

Whilst it can be argued that only a minimal number of users would potentially decode the statue in view of ancient mythology, or even be able to draw upon notions of gender relations of the classical period, it can be assumed that they are familiar with normative representations of gender in contemporary Western culture. Many of the female avatars take the form of traditional Westernised beauty and sexuality that is diametrically opposed and complementary to that of the masculine. Body shapes are typically slim and slender with some emphasis on female curvaceousness, particularly in the emphasis of hips and breasts. Hairstyles are often long and flowing, and the clothing is commonly graceful and glamorous, although some avatars are more distinctly sexual in their attire. There is a noticeable plenitude of female avatars in ball gowns and long flowing dresses which perhaps points to the notion of SL as a realm of fantasy play; it is possible that these avatars are reproducing virtually their owners' versions of a fairy-tale princess, searching for their Prince Charming in a mystical land. Other avatars symbolise more contemporary forms of female sexuality; short skirts and near-exposed breasts are very common. For all the possibilities of SL, and the stated inclusiveness of Apollo, there are overwhelming representations of an ideal figure. As with the masculine, the feminine is generally bounded by the norms and regulations of CL, but exaggerates it through representation, so that the gendered-ness produces an even more pronounced binary of difference between male and female.

Sexuality and difference

Contrary to the perception of the space as an all-welcoming topos of difference, Apollo is arguably a more comfortable space for heterosexual coupling than a space for the expression of divergent sexuality.

Although many avatars visit Apollo as individuals, taking in its serenity and beauty, the island is well equipped for romantic liaisons between avatars. There are various spaces that feature chairs and rugs overlooking beautiful views where avatars are able to activate 'cuddle' pose balls that animate the avatars into non-sexual cuddling. Slow-dancing can also be experienced at the Salsa y Boleros, which is a dance platform overlooking the Apollo coast. Avatars are able to dance with each other by activating pose balls situated on the platform. The island even features a wedding chapel that can be hired out for weddings, which is the major source of funding for Apollo, which has ministers available for services (James Au 2006). According to Ramus Overlord, of *Second Life Traveller*, 'The Lost Gardens are arguably the most romantic sim in the metaverse, from waterfalls, to towers that reach the clouds, not many other places in Second Life have the views and beauty of Apollo'. He also suggests visiting with a loved one: 'being here by yourself can be a bit depressing. The sim is usually jam packed with lovers' (Overlord 2007). Rather than inhibiting hetero-masculine expression, the existence of gay-male sexualities in codified forms does not threaten male heterosexual display, and can instead reinforce it through the expression of patriarchal dominance.

Places such as Apollo are part of the wider matrix of homo-acceptable places that have permeated the globalised and capitalised landscape. Within this landscape, fluid and disjunctured identities are commonplace, allowing greater space for non-heteronormativity and non-cohesiveness with the typical. However, the same forces of globalisation and capitalism that permit such fluidity paradoxically impart a homogenising effect, so that even difference is subjected to the commoditisation of form and expression. Within the subculture of homosexuality and sexual difference on Apollo there is a normative effect acting through commodification on physical perfection and masculinity. According to Warner (as cited in Binnie 1995, p. 185), homosexual culture has been greatly influenced, and is still inseparable from, the creative forces of the capital economy:

> In the lesbian and gay movement, to a much greater degree than in any comparable movement, the institutions of culture-building have been market-mediated: bars, discos, special services, newspapers, magazines, phone lines, resorts, urban commercial districts .... This structural environment has meant that the institutions of queer culture have been dominated by those with capital: typically, middle-class white men.

It is those with capital power who shape the production and administration of space. Although its setting is far removed from the urban city spaces of middle-class Western men, Apollo can be seen as a fantasy island, an escape from the city, but a place that is still bounded by the conventions of the city. The city reproduces the forms and relations that are seen as exciting and marginal yet still subsumed within the dominant discourse.

On Apollo, as well as in the wider SL platform, gender is one of the major aspects of performance that differentiates avatars in their behaviours and interactions with each other and the space. The representations and performances on the island contribute to a discourse of gender that delineates the masculine from the feminine, where it is the masculine that is privileged on Apollo in terms of spatial representation. These representations provide a spatial paradigm for the performance of masculinity within the boundaries of a 'perfected' masculinity of beauty, strength, youth and homosocial interaction, all underscored by ancient Greek symbolism. Performance on the island becomes bounded by these guidelines, so that a perfected form of masculine representation becomes the norm, with the ornamental feminine as its opposite. Edwards argues that 'masculinity is seen to increasingly depend on matters of style, self-presentation and consumption ... [and] is perceived to be increasingly predicated on matters of how men *look* rather than what men *do*' (2006, p. 111). This shows that the contemporary norm of masculine performativity is now more closely aligned to the ancient Greek ideal, although influenced now by the consumerism associated with bodily displays and performances. Corporeal life's commodification of the body provides a potential restriction on Apollo for those who may wish to embody themselves outside of this norm, and thus highlights the impossibility of excising ideology and discourse from virtual spaces.

Avatar performance on Apollo is inscribed by the values of representation: the late capitalist ideal of visual perfection and the commoditised body. Although individuality and expression are esteemed, they are still constrained within the parameters of acceptable performance which usually aligns with the ideologies of success and consumption. The virtual world, due to its reliance on representation, emphasises the body as the key site of expression. Identity is performed as a product of the discourse of Apollo. Male avatars epitomise the Western capitalist ideal of the perfect body, muscular, toned and dominant, whilst the female avatars appear demure in comparison, beautiful but restrained. The avatars that do exhibit difference, such as the 'furries' (half-human, half-animal avatar) or those who exhibit different forms of style (as seen in Figure 3.3) retain a highly gendered appearance. The emphasis

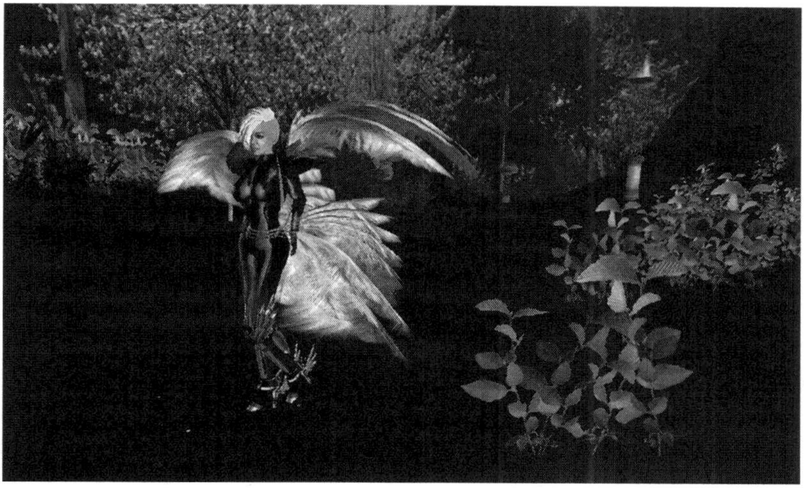

Figure 3.3 Near Hyacinth Valley, Apollo
Source: With permission from Linden Lab and the residents of Second Life.

on gender is inscribed by the space: the male statues, the ancient Greek symbolism, the masculine structures that dominate the island, in what Judith Butler would describe as 'phallogocentrism'. Identity can be conceptualised as a discourse of signifying practices (Jagger 2008, p. 20):

> Identity is rather an effect of signifying practices rooted in regimes of power/knowledge characterized as compulsory heterosexuality and phallogocentrism. As such, it is a matter of social and political regulation rather than any sort of innate property of individuals, or source of agency in a traditional, liberal humanist sense. The political possibilities and agency stem from the inherent repeatability of these signifying practices and the possibility of resignification.

Resignification is key to the potential political power of the virtual. It is the performative aspects of the platform that render it as significant in its application as a site of renegotiated identity, or where the body as subject allows for redefinition.

Conclusion

The environment of Apollo shows how space is produced, conceived and imagined by its creators and users. In the imagining of utopian space, Apollo imagines both the natural environment and the culture of ancient Greece. This reaffirms our view of ancient Greek culture, as well as reconfiguring our relationship with nature and the sublime. The foundation of these significations and our associated understandings of culture and nature are predominantly from a patriarchal perspective. Illustrated through structural signification, the masculine is celebrated, reproducing well-known binaries as a result (masculine–feminine, culture–nature). In contrast, the feminine, and associated feminised sexual practice, is contained as part of the natural environment. Although some subversion can be read into the space of Apollo through the historical narrative of ancient Athens (mostly in reference to historical narratives of homophilic eros), as well as some explicit significations and references to non-normative sexualities and representations, normativity is reaffirmed through the prevailing standards and conduct of its visitors. Heterosexual coupling is normalised through the displays of affection on offer and the hyper-gendered representations of avatars. Non-heterosexuality is covertly displayed through signifiers, unlike the space of Zeus (as highlighted in the next chapter) which offers a more overt connection to gay male sexuality.

4
Masculinity, Mass Consumerism and Subversive Sex

Introduction

Whilst the space of Apollo illustrates a homosocial rather than explicitly homosexual space of masculinity, Zeus is unabashed in its representation of gay male sexuality. Zeus can be considered as a meeting and cruising space for non-heterosexual males, and can also be considered as an extension of gay cruising spaces widely available on the internet. This chapter explores notions of gay masculinities, particularly in reference to the prevailing attitudes of sexual conservatism and the growing acceptance of homosexual coupling within CL. It is in many ways a critique of the growing conservatism of queer culture which, through legitimising certain ways of being, has the effect of delegitimising others as incorrect, immature or deviant. The distancing of mainstream gay movements from alternative notions and practices of sexuality is a disservice to those who engage in sex practices outside of normative monogamous coupling. The heterosexual matrix, with its associations with family, maturity and respectability, has become imprinted onto the lives of gay male identities, proffering wider acceptance within society – it is far easier for the public at large to accept images, ideas and ways of being that mirror a predominantly family-centred cultural ethos. By offering the ideal picture of acceptability and success for gay men, there is opportunity to extend the capitalist marketplace to include this new segmentation, albeit on the ideological bases of youth, whiteness and affluence. Gay men, as seen in images in the public sphere and mass media, are sold to us as physically attractive and masculine, yet asexual and astutely middle-class. It is argued here that the greater purpose of queer collectivity and gay identification in reference to difference has been displaced by a desire to achieve an acceptability of least resistance.

In what appears as a form of assimilation and absorption into the majority, the goal of sexual difference and liberty has been lost. For gay culture, entering into the mainstream has seen desire move underground – coupledom is the new status quo in gay society, whilst promiscuity and free sex are discarded in favour of a desexualised sexuality.

The example of Zeus illustrates a gay male sexuality that exists just below the surface of acceptability. The analysis describes many of the paradoxes of gay male materiality as well as the disjuncture between acceptance and liberty of alternative sexual identities. Located on Gay City Estate, Zeus Gay Club is 'one of the most popular Second Life gay clubs' according to its website, 'having about 700 visitors daily from all over the world' (Ashdene 2011). The site also appears in the Second Life Destination Guide which is often a good indicator of a space's popularity (Gay destinations 2012a). The majority of avatars that visit the site are hyper-masculine in form (female avatars are allowed to visit but they are in the minority) and are consistent with the contemporary ideal of physical attractiveness and hyper-muscularity. Such homogeneity can be conceptualised in terms of the heteronormative and homonormative function of the postmodern gay male; and, superficially, Zeus can be seen to reinforce these notions in its dominant representation of the physically attractive and masculine gay male subject. However, Zeus simultaneously reveals the subversive aspects of gay male sexuality through its construction as a meeting space and cruising space, as well as its explicit marketing of gay male sexual services. Being designated as a sexual space (albeit not overtly sexual in terms of nudity and actual sex – Zeus preludes such encounters), positions the masculinities of the avatars in reference to various moral panics that are seen as 'symptomatic' of an unchecked hyper-masculine sexuality. These issues position Zeus as a contested site of legitimacy in view of tensions regarding masculinity as it applies to sex and sexuality. Furthermore, the site is representative of the tensions between the movements towards an acceptable, albeit conservative, gay male identity as portrayed in Western mass media, and the subversive hyper-sexual and hyper-masculine subject of desire, often viewed as antagonistic to gay male acceptance.

Gay male space

Zeus is predominantly a gay male space, although all avatars are able to access and visit the island. Its positioning as an alternative space is inscribed not only through the type of avatars that frequent the space, but also in the space itself: how it has been designed, its ornamentation, activities, structures and commercial enterprises. As part of the wider personal and political project of gay solidarity, community and agency, Zeus fulfils an important role, as do the number of gay-designated spaces on SL, in providing spaces for identity exploration and the fulfilment of desires. As described by Browne and Bakshi (2011, p. 181), the creation and maintenance of gay spaces has been an important social and political process in the legitimisation of queer communities in opposition to the heterosexual norm:

> 'Straight space' can thus be seen as asexual for mainly straight people and heterosexual for LGBT people. Such spaces are also heteronormalised, by this we mean that heterosexuality is formed through particular gender/sexed binaries within normative male/female dichotomies.

Corporeal LGBTI spaces have been a response to heteronormativity, providing sites of safety, solidarity and community, where individuals can develop and perform queer identities outside their often assumed heterosexual identities which are performed in the everyday public and private spaces. As an extension of the corporeal, virtual sites like Zeus can serve as an extension to the project of challenging assumed heterosexuality by providing an arena for sexual and identity exploration.

Authors Browne and Bakshi, however, suggest that the binary between straight and gay is currently being challenged through both the colonisation by queers of heterosexual spaces (the gaying of straight space), and the colonisation by heterosexuals of queer space (the heterosexualising of queer space, which can also be described as infiltration – see Chapter 5) (2011, p. 181). The dilution of spatial conformity through sexual difference can be seen as a positive for contemporary society by making the queer subject more visible in that the queer subject is not simply reduced to a locus legitimated only by queer space and through other queer identities. In a poststructuralist sense, queer identity becomes unnecessary due to the inclusivity of post-urban as well as virtual space. Here, there is limited need for oppositional identity assumed around queerness (where queer is assumed as politically reactive). It is arguable though, as discussed in Chapter 5 regarding the female-only space of Eden, that this position is premature because there remain continuing threats to queer identities that cannot simply be overlooked in reference to poststructuralist critiques. The infiltration (or de-gaying) of queer spaces can have the effect of marginalising specific subject groups that do not conform to notions of acceptable identities.

Spaces designated as queer also serve an important function in the coming-out of gay males, as well as playing a key role in the formulation of identity, offering 'opportunities for action and coalescence' and 'allowing for the showcasing of diverse sexuality.' (Pullen 2010a, p. 1) There is indeed an on-going necessity for such spaces to exist, as described by Christopher Pullen who recounts the murder of a gay Californian youth (Lawrence King) in 2008 due to his sexuality. According to Pullen, the LGBTI publication *The Advocate* flags the issue as one of potential over-confidence for the queer community 'which might inappropriately support a youth to express sexual diversity at such an early age, and become vulnerable' (2010b, p. 17). Happening relatively recently, King's murder highlights the necessity for queer individuals (particularly youth and other susceptible demographics) to have spaces safe for individual expression. It could be argued that for gay males in Western societies, this issue may be more pressing due to the potential for violence at the hands of heterosexual males. According to the FBI's Criminal Justice Services Information Division, in 2009 there were 1,482 victims of a sexually oriented hate-crime reported in the US, of which 55.1 per cent were 'victims because of an offender's anti-male homosexual bias' (where specific anti-female homosexual bias accounted for 15.3 per cent) (Justice 2009).[1] Contrary to the theoretical assumption that the categorisation of 'queer space' may have lost its usefulness, there certainly remains a need for there to be such spaces in society given the very real dangers that continue to threaten queer identities, regardless of growing acceptance of gay sexualities in the wider community, and as offered in the mass media.

Although mass media portrayals are often legitimising for queer identities, they are, at the same time, limiting in the type of identities that are legitimised. And, whilst such representation is generally a positive for minority identities in that they achieve acknowledgement and visibility, the case of Lawrence King highlights the dangers associated with the perceived acceptance of these representations in the community. Specific spaces, set aside from general public space, are both necessary as enclaves of safety for queer identities, as well as serving as areas in which to participate and interact outside of the established norms of identity and conduct dictated by normative representations. As argued in this chapter, there is a conservatism, often associated with the wider acceptance of gay identity, that can act in opposition to the liberalism of queer (specifically gay male) identification; and this conservatism sets up a binary for modes of conduct that are acceptable (and therefore legitimate) in opposition to those that are deviant (and ultimately delegitimising). In analysing queer representations on television, Guillermo Avila-Saavedra (2009, p. 5) states:

> The presence of homosexual characters in American television would seem to imply an endorsement of a liberal agenda of tolerance and inclusion of alternative lifestyles and sexual orientations. However, the perceived progressiveness of gays' sudden appearance on American television could be undermined if it responds to traditional norms of social relations.

Furthering this argument, Paul Robinson suggests that 'the emergence of gay conservatism as a political and intellectual force is arguably the most important new development in the gay world' (2005, p. 1), and that this ultimately undermines gender and sexual diversity. The gay male in mass media depictions has become homogenised and suitable for mass consumption through mimicking the heterosexual standard, whereby 'Homosexual images are presented in a way acceptable for heterosexual audiences by reinforcing traditional values like family, monogamy and stability' (Avila-Saavedra 2009, p. 8). Moreover, consumption and class assertion have also become characteristic tropes of gay identity through mass communication channels, so that this once subversive identity is now easily subsumed into the heterosexual matrix of acceptability and hegemonic control.

The site of Zeus illustrates some of the contradictions that abound within the politics and expression of gay male identities. Whilst it is a space that allows for the more subversive aspects of gay male sexuality, it is simultaneously conformist to notions of gender (heteronormativity) as well as materialist notions of homosexuality. The heteronormative is expressed through the representations of masculinity and femininity as being normatively assigned to both male and female avatars. Male avatars, although often appearing shirtless and displaying a gay male symbolic appearance (shown in tight-fitting clothing and colourful attire), are also highly masculine, generally being tall and muscular (see Figure 4.1 below). Females, conversely, epitomise femininity, and often appear slight by comparison and are dressed in tight clothes that emphasise curvaceous figures rather than the hard muscularity of male avatars. The female–male binary is upheld with strong signifiers of femininity and masculinity as attached to biological sex and gender categories. In addition to gender norms, the space also represents aspects of homonormativity which can be described as 'the construction of social norms that include lesbians and gay men on the condition that they conform to individualist and consumerist economic values and lead sexual lives that mirror the norms of heteronormativity (e.g. long-term, monogamous relationships within specific gender norms)' (Browne & Bakshi 2011, p. 181). This is expressed on Zeus as an attachment to the mass consumerist depiction of the homosexual male: white, middle-class, masculine and appearance-conscious, but makes an important departure from it in reference to sexual conformity and the heterosexual standard of coupling and monogamy. The moderate/mature rating of the region, as well as its appeal as a niche space, allow it to represent and promote a gay male sexuality that is in a large part disregarded in mass media portrayals. Homosexual sexuality and bonding are now arguably considered safe in mainstream Western culture in a de-sexualised form, largely due to the reification of queer to a commoditised and conservative identity. The supposed rampant sexual desire and promiscuity that have historically been seen as stereotypical traits of the gay male subject are now contained as inappropriate extensions of hyper-masculinity acted through deviance.

The SL region of Zeus illustrates the tensions between a representational space of legitimate gay male identity conforming to the

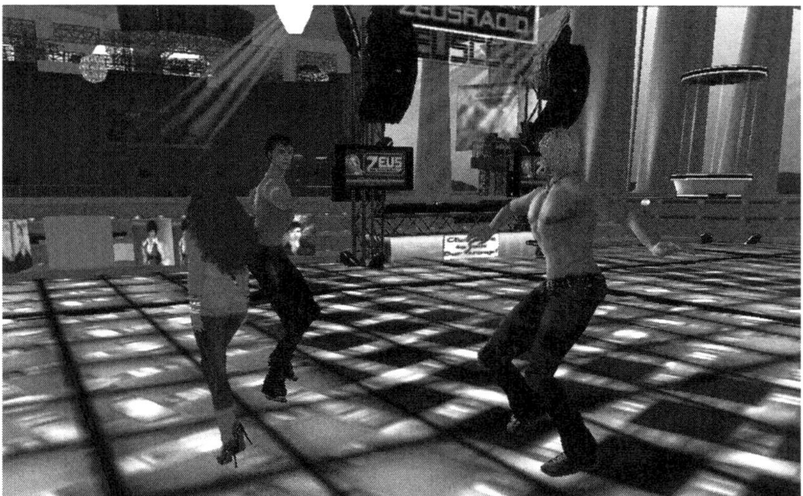

Figure 4.1 Dance-floor, Zeus
Source: With permission from Linden Lab and the residents of Second Life.

commoditised expression of the body (capitalist masculinity) and one which allows for behaviours contrary to the conservative norm of homosexual conduct (promiscuity and hyper-sexuality). Michael Warner describes the tension between the two worlds as a hierarchical power struggle. Utilising Erving Goffman's terms of 'stigmaphiles' (those who are stigmatised) and 'stigmaphobes' (those who are normal), he states:

> The stigmaphile space is where we find a commonality with those who suffer from stigma, and in this alternative realm learn to value the very things the rest of the world despises – not just because the world despises them, but because the world's pseudo-morality is a phobic and inauthentic way of life. The stigmaphobe world is the dominant culture, where conformity is ensured through fear of stigma. (2000, p. 43)

For the most part the stigmaphobes have been winning the battle for morality against stigmatised deviance, so that even in the progression of queer acceptance, there is a ubiquitous moralising force regarding acceptable sexual conduct. Spaces of deviance are quickly consigned to the margins and often renounced by the interest groups that should act in their political interest; it is no coincidence that gay movements, in all their successes, have become desexualised in their agenda. The concern is that the homogenising force of homonormativity may marginalise the diversity of identity and expression within the community, thereby becoming a stigmaphobe world for those it should embrace. Of particular interest for the rest of this chapter is how this force works to organise ideas about strict gender representations and divisions, the control of sexual conduct, and moral panics regarding the deviant.

It is important to position this discussion in its political and social context, given that SL draws heavily upon European and US norms. Although SL is a global medium, user statistics show that participation is located most highly in Europe and North America, with Germany and the US having the highest number of national users as of 2007 (Europeans predominate as Second Life users: study 2007); recent statistics have proven extremely difficult to come by, possibly due to the reluctance of SL's owners to publish declining user numbers. Although Europeans participate broadly in SL, the political and social context is most closely aligned to the US due to its deregulated economic system (it also relies heavily on US currency as the basis for SL's Linden dollar). The platform is reflective of neo-liberalism in that the economic system takes prominence over activity whereby deviance is arguably tolerated. The premise behind this attitude is that tolerance of deviance, as well as tolerance of problems, is acceptable 'as long as said tolerance results in improvements in bottom-line profits' (Marzullo 2011, p. 762). Neo-liberalism works alongside neo-conservatism through the reinforcement of dominant tropes of gender, sexuality and identity that are played out through debates on morality and dominant value systems. According to Connell, this functions ideologically 'as a form of masculinity politics largely because of the powerful role of the state in the gender order' and, in so doing welds 'exemplary bodies to entrepreneurial culture' (2005, p. 1817). This allows for individualism, insofar as the individual conforms to the dominant system, which ultimately produces a homogenised version of gayness and homosexuality largely understood through the framing of hetero-patriarchal politics. Sharif Mowlabocus comments that the homosexual has been turned into a 'marketable commodity', yet is also 'sanitised and safe' to conform to family values (2008, p. 428).

The performance of masculinity

Notions of masculinity are as important to the homosexual subject as the heterosexual subject, and traverse corporeal and virtual spaces. It can be argued that anywhere a body, or an individual, is represented it takes place through an adherence with, or opposition to, masculinity. In apposition with historical patriarchy, identity is often formed through an essentialised, yet ubiquitous, framing of maleness through masculinity. Its modern conception has developed from classical Greek and Roman societies through the Enlightenment and Industrial societies to its modern and postmodern configurations. What has remained throughout the history of Western masculinities is the power and access of the male subject: as a full citizen (classical societies), to own land (aristocracy), to participate in the marketplace (modernist-capitalist societies), and to control women (both sexually and economically throughout history). David Buchbinder (1998, p. 43) explains such modes of differentiation as spoken through power:

> [P]atriarchy ranks and thus creates power differentials even among those who it centralises. In this way, differences among individual men, such as age, physical size and strength, class, wealth, social or political clout, sexual activity or hyperactivity – even penis size – and so on are invested with varying degrees of patriarchal power. It is

> the aggregation of these elements and their investment of power by patriarchy that we recognise as 'masculinity' or the lack of it.

Such mastery has become an expectation, a necessary performance to reaffirm individual membership as a man, even through the changing dynamics of gender relations and feminist assertions of contemporary life. It is perhaps no surprise that the control and strength associated with masculinity should turn inward in its response to changing social practices that seek to moderate its control. Late-capitalist masculine expression often articulates its qualities through a mastery of the marketplace, evident in conspicuous consumption practices, as well as a self-mastery of the body's physical form.

The contemporary male body can be said to have returned to the classical ideal of the muscular and beautiful male body, although importantly often 'forgetting' the homosexual undertones of this period. The case study of the SL space Zeus exemplifies this through both its name and its associated mythology. King of all Greek gods, the mythical figure Zeus symbolises the ideal male. He was physically intimidating, astute and controlling of the world around him. Strong and commanding, he was the moral and judicial leader of ancient Greece, as well as being the protector of Greeks and their societies. Moreover, he was the king of the heavens who oversaw the conduct of all other gods. It is these ideas of perfection, control, mastery and competence that are heavily evoked on SL's Zeus through the structural environment and bodily representation. Existing somewhere between the heavens and mortal civilisation, the myth of Zeus represents a half-way point: a journey from the ordinary world to a place where immortality and the gods are accessible. In embodying perfection, one's avatar can access the gods: he (or she) can fly, teleport, and change the weather and environment at an instant through the keyboard. Additionally, hedonistic, particularly sexually promiscuous, behaviour as perpetuated in Greek myth is permitted. Although more inclined towards females, Zeus did have the occasional male lover, but perhaps more importantly it is his unflinching sexual domination of his desired object that resonates through mythological accounts. As the supreme god of the heavens, Zeus could enact his desires when and where he chose to do so, even utilising a chameleon-like ability to 'trick' his desired objects of affection into sexual liaisons. Parallels with the chameleon qualities of SL are evident here in the possibility to embody any human, or even creature, in the visual sense for the possibilities of sex or otherwise. The myth of Zeus not only represents physical and metaphysical mastery of the environment, but also sexual domination. As noted by Michael Kimmel, and as displayed through the representations on Apollo, as well as Zeus, this mastery and domination are principally displays for the gaze of other men: 'Masculinity is a *homosocial* enactment' (2001, p. 275).

Contemporary society has in many ways returned to this ancient ideal through the championing of an often unachievable physical form. The active development and display of the male body draw heavily upon the classical form and mastery of physical beauty and prowess as epitomised in the imagery of Zeus. This unachievable body, the pinnacle of masculinity, shows its influence through the popular pursuits of bodybuilding and gym work. In his analysis of the role of bodybuilding, including the paradox between masculinity and the male gaze, Mark Simpson describes the unease which led to the acceptance of the sport and the continuing tensions that threaten to position the development, active judgement and tacit approval of the near-naked male form as deviant. He argues that:

> while the appropriation of bodybuilding to buttress the image of an increasingly unstable masculinity appears to have been phenomenally successful, it is itself inherently unstable, its unsavoury past always threatening to gatecrash its new-found respectability and expose masculinity's own scandalous secrets. (1994, p. 27)

There is a persistent paradox of the hyper-masculine form in its desperation for heterosexual affirmation that reaffirms not only the historical positioning of ancient sexuality, but also the contemporary approval of the male gaze looking upon a desired male form. As Simpson suggests, although the mainstream culture has been resolute in framing the male body as fervently heterosexual, 'it cannot erase the fact that its use as a way of socialising males into heterosexuality *is utterly predicated upon its homoerotic appeal*' (1994, p. 29; emphasis in the original). This shows the tenuousness by which modern conceptions of masculinity are based on notions of the homo-gaze (appearance and performance for other males). Although underscored by notions of homoeroticism, this gaze is rationalised through a heterosexual identification that contains and controls homosexual arousal.

Bodybuilding may illustrate mastery of the physique in the corporeal world, in which it could be argued that power and strength are the major reasons for investing in the body over simply being gazed upon. However, there is no corollary in virtual space to the notion of physical strength. What remains within the virtual performance is the gaze; one encourages others to look upon the body, for it is only the body appearance that holds agency in unfamiliar relationships. Heterosexual, as well as homosexual, males need to present themselves with particular reference to the visual. The virtual environment renders masculinity entirely performative, mainly as a representation of an avatar's appearance. On Zeus the dance-floor acts as the space of spectacle where the body can be displayed in its full performativity. Although dance is commonly associated with femininity and the female body as spectacle, on Zeus the spectacle is the male body. However, it is important to note that the male body is not feminised; it remains a spectacle of masculine beauty. And whilst the avatars that visit Zeus display gay male signification they do not fit passive, or effeminate, identification. Rather, they display masculine beauty as typified in mass culture. The reliance on gay beauty signifiers is perhaps in a response to the primacy of the visual. In researching gay male chat-rooms, John Edward Campbell (2004) found groups re-writing beauty in response to these mass depictions, and thus reworking depictions of desire through participation and sharing – a phenomenon that is absent on Zeus.

Bodies on Zeus emphasise the masculine qualities of the male, where the body falls back on signifiers of action and physical competence, rather than flamboyance or submissiveness. Avatars are proportioned to emphasise height, bulk and strength. They are sexualised in their dance moves, but these movements inscribe action and muscular display – they are far more like the bodies seen at circuit parties than at pop events (Peterson 2011). The SL scripts that animate the avatars are set at default movements for gender, so that male avatars walk with larger gaits than females and with more exaggerated movements. This can be changed within the SL environment, but to do so requires either Linden dollars to invest in animating an avatar in a different way to the default, or the ability to master the technology itself. Whilst it is possible to re-animate an avatar, gender normativity in reference to movement is generally maintained alongside physical appearance. The only major element differentiating the male avatars on Zeus (presumably homosexual) to others in SL's non-gay environments is their clothing, or lack thereof. Gay expression abounds in the avatars' attire, sporting revealing cuts and materials as well as signifiers such as cowboy hats and BDSM wear. It could be argued here that such clothing and appearance options feminise the body in its invitation to be looked upon. However, there is a complexity within the adoption of a subject-object position that cannot strictly be explained in terms of femininity. Whilst language often resorts to explaining behaviours through the female/male binary, such as in the assertion that a male who assumes the 'bottom' position is passive or somewhat girl-like (similar to the Freudian conception of inversion), these categories are too simplistic to evaluate the performances as seen on Zeus. The avatars are undeniably male (as a presumption of an underlying biological sex category), but they are also masculine. What they perform, much more so than in heterosexual culture (although this is becoming more commonplace), is a desire to be looked at, applauded and desired in their masculine beauty. Their clothing and desire to be seen are an extension of the narcissistic self, not in reference to the Freudian-pathological conception of homosexuality, but to the concept of late-capitalist masculinity itself.

Such narcissism on Zeus is predicated upon the heteronormative consumerist ideals of status and power. Hyper-capitalism has seen the expression of this ideal become centred on the body, whereby 'men are defined now more than ever through their consumption, sexuality and physical appearance' (Brubaker & Johnson 2008, p. 131). Whilst some scholars have offered that this trend has shifted gender distinctions (Gill, Henwood & McLean 2005), it can be argued instead that the aestheticisation of the male body is a new type of masculinity that both reinscribes and reinforces masculine discourse and hegemony. Rather than being feminised, new masculinities assert a phallocentrism and differentiation from the female through a gender identity performed for men. Hardness, in its physical (muscular and phallic) and psychological (mental and emotional toughness) embodiment, is a state of expression for men as a performance of mastery and power. The female object is only a referent: as something of comparison. Due to the contemporary focus on the male body as object, the opposition to the female becomes even more important for the assertion of masculinity, through hyper-masculinity. There is some suggestion 'that men are responding to this crisis through increased social space for men's bodies and emphasis on size, strength and violence as valued components of masculinity in popular culture' (Brubaker & Johnson 2008, p. 132). Gay men have also bought into this notion of success subsumed under hyper-masculinity. In what can be seen as a refutation of the gay male as 'being in some way or other "like" a woman, fey, effeminate, sensitive, camp' (Dyer 2002, p. 5), there has been a culture of virile queer guys in images, fashion and pornography post-1960s. As explained by Dyer, even in the absence of women, gay men's sexuality is, at least in part, defined through either an affinity with, or opposition to, notions of femaleness and femininity. The commodified body, and the gaze it invites, are reinscribed as masculine performances; by men and for men in a rejection of the feminine.

As a required oppositional subject position, the feminine is necessary. Within the space of Zeus, femininity functions as a representation of heteronormative female avatars within the space as well as being an inscribed opposition to the masculine. As within corporeal public and private space, the presence of females has the effect of legitimising gay male sexuality in reference to the feared dangers and deviance of an unchecked hyper-masculinity assumed to be both uncivilised and uncontrollable. Male sexuality is seemingly moderated, more civilised and less threatening. Sender remarks that the 'constraints historically placed on women's sexual expression have been naturalized to appear as a consequence of (particularly white, upper-class) women's low sexual drive, prudery, attachment to monogamy, and sexual vulnerability' (2003, p. 352). The presence of heteronormative females on Zeus works to not only moderate the notion of rampant gay male sexuality, but to further entrench the idea of the (white, middle-class) female body as asexual, through their non-participation in an active sexual desire. Their presence can be viewed as ornamental, adding to the glamour and sophistication of the scene, as well as its cosmopolitanism (Binnie & Skeggs 2004). Women in gay spaces are meant to be beautiful but not active. This is reflective of the presence of heterosexual females in CL gay clubs and bars, which are often considered safer than heterosexual spaces for women to socialise, protected from the advances of heterosexual male desire.

There is an undercurrent of not wanting to be perceived as feminine that pervades Zeus: the effeminate male is peculiarly absent. Although there are numerous signifiers of gayness throughout the space (the club environment, the colours, the flamboyant, and often absent, clothing), there is an obvious refusal to represent the self as effeminate. As a fantasy world, Zeus replicates the sexual desires and visions of the male body as mass produced in the gay mainstream. Bodies are tall, muscular and hard; desire and sexual liaisons are predicated upon this masculinity, where the female-object role as per heteronormativity is expressed ultimately through sex itself. In a mirror version of the theory of lesbian 'inversion' (where the more masculine subject assumes the male position), the 'male' role is activated through the sex act: that who does the penetrating remains male whilst the one receiving takes on the feminine position. Like Halberstam's reasoning of lesbian sexuality (understood as separate from gender binaries and heteronormativity), so too must gay male sexuality not simply be reduced to an all too essentialised gender mimicry. The problem with the supposed dyad is in its reduction of the masculine to ideas of domination as it resonates with heterosexual categorisation. Tim Edwards argues that the issue of domination within gay male pornography is 'the explicit oppression of certain types of masculinity in these images' (1994, p. 88) whereby the less dominant, and therefore less masculine male, is subjected to the more dominant male's desires. He explains that this positioning of the lesser can be viewed through the broader lens of societal and cultural domination practices that seek to subjugate gender diversity and non-conformity (Edwards 1994, p. 88):

> More significantly still. This process is then juxtaposed with the oppression of male same-sex sexuality at a societal level which is constantly played upon in the pornography as the passive or powerless partner is always perceived as more primarily homosexual while the active or powerful partner's sexuality is perceived to transgress from straight to gay and, in addition, the gay consumer of the pornography is led to identify with the pre-given gay partner in eroticising the transgressive straight to gay partner due to the camera's construction of the passive partner as subject and the active partner as object.

There is an obvious corollary here with any visual media that invites the passive gaze, and therefore positions the viewer as the homosexual recipient of arousal. Virtual 3D environments, however, may challenge this phenomenon due to passivity being replaced with active participation: if one can embody the more powerful position in a scene that is reminiscent of pornography then the associated power assumptions (and the links between femininity and homosexuality, particularly taking the 'bottom' position) can be circumvented. However, problems remain regarding the conventions of femininity and masculinity and the location of desire and agency as dictated through the norms of culture and power relations. If, within Zeus, a male wishes to embody the powerful masculine position, he does so at the expense of another. Masculinity remains caught up in its presumptions: to be masculine means to position the object-of-desire as feminine and a lesser agent in sexual activity. On a broader level, it reinforces the subjugation of the feminine male as too-close-to-female, and therefore external to the privileges of masculinity.

Commodification

Contemporary discourses of masculine power are often enacted through consumerism. Susan Alexander argues that since the 1950s man has adopted a new form of masculinity that is a consumer product that 'rests on one's outward appearance rather than on the traditional male role of production' (2003, p. 551). Mackinnon argues that this change in masculinity can be attributed to contemporary capitalism's incompatibility with the patriarchal division of traditional gendered behaviours (1997, p. 26):

> When it no longer, as it were, suits capitalism that there be two antithetical, mutually exclusive genders, the patriarchal demand for belief in them is undermined. If contemporary capitalism needs, in addition to security of production, a technology of consumption together with the legitimation of desire, it is fair comment that the differentiation of bodies by sex is increasingly irrelevant.

This is particularly apparent in gay male sexualities as part of the urban consumer-class consciousness enacted through both the ability to spend on the body as well as discretionary spending on lifestyle products and services (Altman 1996; Sender 2003, 2004). The space of Zeus reflects this, having a large shopping precinct where one can buy items for bodily adornment as well as those for a leisure-filled lifestyle in-world, such as lighting rigs and DJ equipment. Outside of this precinct, and in various spaces throughout the region, there are many billboards advertising various products and services tailored primarily towards the gay male market. The emphasis is on the male body as beautiful in both its sexuality and masculinity, with the important departure from the classical Greek idealism of male beauty being the commodification of that idealised representation (Figure 4.2).

The body is not simply a product of hard work and physical mastery, as epitomised in ancient Greece, but something that can be bought: a commodity related to the mastery of capitalism and market savvy. This is representative of the segmentation and categorisation of the gay market in Western capitalist societies, where the gay male market has been specifically identified as 'a sufficiently large and profitable group to warrant marketers' attention' (Sender 2004, p. 1). This has attracted a large number of companies to appeal to gay male consumers over the social and political complexities that are present within society. The gay male market is often considered to have a large disposable income

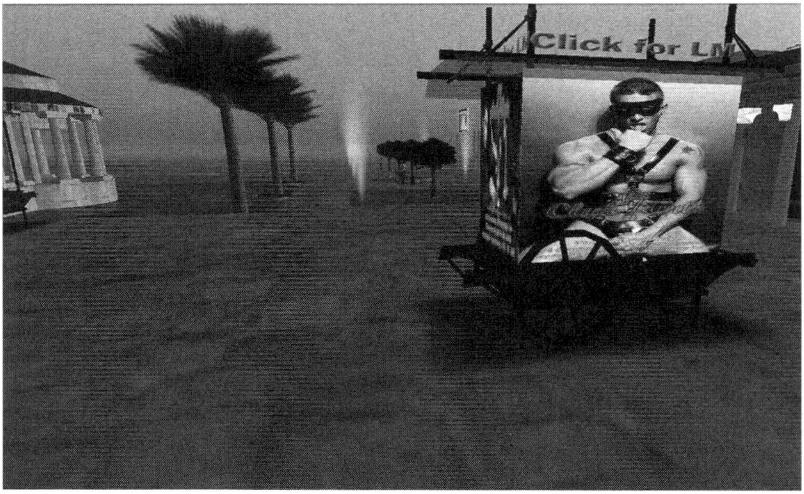

Figure 4.2 Advertising board, Zeus
Source: With permission from Linden Lab and the residents of Second Life.

and a high investment in personal appearance and status, prompting particular attention from producers of high-end lifestyle products and purveyors of conspicuous consumables. Katherine Sender (2004, p. 5) discusses the idea of discursive categorisation in reference to marketing and advertising in the US:

> [T]he gay community ... is not a pre-existing entity that marketers simply appeal to, but is a construction, an imagined community formed not only through political activism but through an increasingly sophisticated, commercially supported, national media.

Such discourses reinforce an identification of gay-identifying men to social expectations of what it means to be a gay male in Western culture. It is important to note here that such market segmentation helps to produce sub-categories of gay male sexuality. As opposed to presenting a cohesive unified whole, gay male sexualities are segmented along class and race divisions, as well as political allegiances, further entrenching norms into the pre-existing power structures of white middle-class masculinity. That whiteness is assumed is significant in the construction of commercial gay identities where black, Asian and other ethnicities are most often marginalised through their absence. This is particularly interesting given the position black males often hold vis-à-vis hyper-masculinity and hyper-sexuality. Such exclusion reaffirms 'the model of physical perfection embodied in Classical Greek sculpture (which) serves as the mythological origin of the ethnocentric fantasy that there was only one "race" of human beings who represented what was good and true and beautiful' (Mercer 1991, p. 192). Furthermore, the pervasive whiteness of the avatars signifies the culture of technology and the digital where there is a presumption of whiteness, if not at the very least as shown by Andil Gosine (2007) in a study of race in Gay.com, and the privileges afforded such whiteness.

Although much of mainstream media seeks to desexualise the gay market in order for it to be palatable for heterosexual consumers, marketing for gay men within gay media and spaces is often highly sexualised, serving a market outside of the acceptable conservative norm. Zeus is a space for looking, and for selling oneself as a sexually available and desirable gay male body, as evidenced by the number of sexual advertisements found around Zeus. These depictions are anathema to what Sender describes as the publicly acceptable face of homosexuality, whereby '[a]nxieties about gay men's sexuality, embodied in the two stereotypes of the hypersexual, predatory, possibly paedophilic gay man and the promiscuous AIDS victim, have shaped the constitution of the ideal gay consumer' (2003, p. 332). This 'ideal' consumer is seen as either non-sexual or barely sexual, where 'the 'charmed' (or at least less abject) manifestations of homosexuality have become the public face of gayness' (Sender 2003, p. 333). SL gay spaces such as *Zeus* represent the more subversive side of gayness that is often rejected in dominant media representations, whilst still reinforcing many of the dominant ideas of male consumerism, especially relating to the lack of difference in class, race and age depictions. This subversive market has been re-produced and refined in a way that draws upon the dominant discourses of male homosexuality, ultimately reinforcing dominant representations of male beauty, masculinity and promiscuity that exist as a counter-public to the now acceptable face of homosexuality.

Although the ideal cultivated image of the gay male is of the 'affluent, white, male, thirty something, gender-conforming, and sexually discreet' (Sender 2003, p. 335), spaces in SL are not bounded by the same appeals to heterosexual conformity and conservatism as gay media in the mass public sphere. Spaces like Zeus, which tailor to this counter-public, are at the boundary of the public/private sphere. Although the space is accessible to anyone with the hardware and software capacity to enter the world, it has the veneer of privacy, due mostly to the anonymity of identity. This, coupled with the number of regular users of the site, means that on a global scale spaces such as Zeus remain relatively guarded from the public masses. Such tolerance of the wider gay market as well as the niche market of gay sex and subversion points to a tolerance aligned with economic determinism whereby such identities are permissible in reference to their usefulness in the marketplace. Rob Cover (2006, p. 296) explains that:

> This 'repressive tolerance' motif indicates a shift in viewing exploitation, along with many marxian traditions, in the sphere of production and indicates that exploitative repression occurs also in the categorization of a market identity whereby subjects are encouraged to consume in order to 'fulfil' their ostensible identity.

However, this usefulness is generally predicated alongside the prevailing norms of the heterosexual matrix so that ideologies and discourses are reaffirmed within the economic structure as well as the cultural. Through the intense commodification of the body, the gay male on Zeus is inscribed as a master consumer. The major departure from mass media representations of gayness is the sexuality inscribed into the space whereby the body is seen as the site of extant desire. Crossing over from the more public face of homosexuality, the shops and representations of Zeus illustrate the marketplace extending into the subversive. As with pornography and sex shops, the non-conformist is allowed to practise deviance dependent on their economic agency.

Although the space of Zeus rejects the conservatism of mainstream media depictions of gayness, it reinscribes many dominant and historical signifiers: the gay male as beautiful, affluent and promiscuous, as well as predominantly white. It can be seen as part of the greater industry of internet communication technologies that market within the industry of social connections for the homosexual subculture, particularly that of the sexually subversive. Parallels can be made to sites such as Gaydar (an internet site used by gay men looking to find other gay men). Research into the site has found that, although it is not a sexually explicit site (not explicitly pornographic), it is used for introductions as well as providing an escort service (however, it is also used by men for various purposes including finding friends and creating social networks) (Light et al., p. 307). Similarly, the preferred type of gay man is inscribed through its sign-up process which is further reinforced through marketing and branding. Research conducted by Light,

Fletcher and Adam (2008, p. 306) discusses the preferred advertisers of Qsoft (owners of Gaydar):

> QSoft continuously enrols advertisers whose specific visual and textual representations of sexuality conform to and confirm specific meanings of attractiveness and affluence. The enrolment necessarily commodifies the sexuality of individual members and in doing so encourages these members to reflect on the extent to which they reveal themselves and their identities.

Like Gaydar, Zeus inscribes the preference for the ideal gay man as a sexually available, masculine, muscular and affluent subject who is discerning about others. There is a reinforcement of masculine power through the representations depicted by advertisers as well as the prevailing norms of users. Through the intense commodification of the body, the gay male on Zeus is inscribed as a master consumer, eliciting homonormativity through consumption and mimicking the practices allowed in the heteronormative conservative marketplace. The major departure from mass media representations of gayness is the hyper-sexuality inscribed into the spaces, images and subject and object positions of the avatars. Whilst avatars have been successfully defined by the prevailing attitudes of masculinity and the beautiful consumer body, subversion comes through the sexual practice of deviance.

Moral panics of subversive sex

The night-club space is significant in gay culture, having deep historical and cultural connections with subversive sex practices (as defined by the mass public). With roots in the bath-houses and bars of the 1970s, such spaces are often enclaves for alternative sexual proclivities and practice, being 'safe' from judgement, harassment and often violence (Bérubé 2008). The connections to subversive sex remain even in the development of progressive sexual politics that have seen greater visibility and acceptance of non-heterosexual coupling. Through digital technology, the cultures and practices of non-normative sex often replicate the signifying practices of the corporeal, so that chat-rooms (in text-based communication) and 3-D virtual environments resemble the bath-houses, bars, night-clubs, tea-rooms and cottages of the physical world. In researching casual sex in a university environment, premised by activity on the site Uni_cock, Sharif Mowlabocus describes the connection of technology with illegitimate behaviour and the extension of digital culture from and to the physical (2008, p. 434). Similarly, Zeus is an extension of these cottaging, tea-room or tacit hook-up practices, where liaisons are acted out through the screen, rather than using the screen as a precursor to corporeal sex. Still, such 'lifestyle choices' remain outside of the legitimate sphere of sexual practice, particularly given the anonymity and limited temporality of such practices. Furthermore, there are specific connotations of masculinity here, where the anonymity and fleeting nature of sex resonates with an unfettered masculine sexuality that operates outside of 'gay' insofar as gay is understood in the mass culture. The anonymity and fleetingness of such encounters plays into a fantasy of masculine sexuality, whereby one is not necessarily considered gay but engages in dangerous or deviant sex, such as in the 'great dark man', 'rough trade' or sex with heterosexual men.

Other practices of deviance are also reified and expanded through the internet, such as pornography and barebacking (sex without a condom). Finlayson (2012) suggests that 'For most gay men the politics of sex are subsidiary to their need for emotional and sexual intimacy. A condom coldly reminds them of HIV status, whereas latex-free sex is seen (and felt) as intensely liberating'. Furthermore, in a 2003 study by Halkitis et al., who researched the practice of barebacking among gay and bisexual identified men in New York, 30.7 per cent of respondents indicated that they agreed with the statement that 'barebacking is popular because of the internet'; 45.5 per cent of the 448 men surveyed who were familiar with the term 'barebacking' indicated that they had participated in the practice in the three months prior to the survey) (2003, p. 353). The major reason for implicating the internet in the popularity of barebacking was due to the ease of finding others anonymously and in reference to hook-ups on the internet for casual sex. Importantly, such research also acknowledges the issue of risk-taking behaviours that can be associated with bars, clubbing, alcohol and the rise of crystal-meth in the late 1990s (Dean 2009, p. 102).

Such risk-taking, as well as the gay pornographic culture of barebacking that privileges 'raw flesh' over safe sex, reinscribes a type of hyper-masculinity into gay culture, suggesting that 'the internet might be *productive* (not just reflective) of emerging sexual cultures' (Dowsett et al., 2008, p. 122). Halkitis et al. suggest that 'gay and bisexual men perceived numerous psychological and emotional benefits associated with barebacking, including but not limited to feelings of connectedness, intimacy, and masculinity' (2003, p. 355). Similarly, Dowsett et al. found in their study of barebacking and masculinity on US internet sites that masculinity was central to the practice, albeit a different manifestation of the masculine from what is understood through the heteronormative (2008, p. 125):

> We found a kind of masculinity in which that articulation of sexuality and gender was exercised in different and unique ways, neither as simply derivative and a replica of heterosexual men, nor as a superficial text and performance through which the palimpsest of 'real men' could be seen.

Dean suggests that barebacking is a subculture unto itself, 'with its own distinct identities, rituals, and iconography' (2009, p. 104). Sex-based cultures embody a different sort of masculinity through practices that reify masculinity via sexuality. Whilst signification can simultaneously connote 'gay' as well as 'masculine' from extraneous observation, the practices performed as inherent to the culture are far more complex in terms of gendered and sexual performance. A specific problem associated with commercialised and subversive spaces and identities is how to position the self and sexuality relative to the normative culture. Zeus displays these tensions with meaning derived from the commercialised body as well as from the subversive body, which work upon each other to inscribe various sign-values of representation generally iterated through desire.

In view of Western conservatism, Zeus can be framed as a seedy underbelly of illegitimate behaviour. With particular reference to phallocentrism, the homosexual male is unchecked in its deviance and unstable as an individual agent of desire, signifying the 'masculine libido incarnate, the dangerous antithesis of family and community' (Stacey 2004, p. 181). The internet, and its associated connotations of deviance, works to produce spaces considered anathema to the 'normal' and controlled functioning of society (Sender 2003, p. 333):

> Control over the body has long been considered essential to producing an orderly work force, a docile populace, a passive law-abiding citizenry. Just consider how many actual laws are on the books regulating *how* bodies may be seen and what parts may not, *what* you may do with your body in public *and* in private, and it begins to make more sense that the out-of-control, unmannerly body is precisely what threatens the orderly operation of the status quo.

Outside of the mainstream acceptance of heteronormativity and homonormativity, spaces like Zeus demonstrate the supposed dangers of hyper-consumerism and the overly narcissistic attention to the body and its pleasures. However, as put forward by Judith Stacey, such spaces can position male homosexual culture as 'a potent source of oppositional values and cultural resistance' (2004, p. 182). This subculture may well pose a threat to the established order of domesticity and the politics of coupledom, but should not be viewed as a threat on an individual level as a form of deviance and incivility. What is potentially antagonistic to the social order, below the superficial arguments regarding legitimate sexual behaviour (including promiscuity as an assault on monogamy and coupledom, and rampant hyper-sexuality) is the very notion of masculinity itself.

The undermining of constructed gendered positions may well be the source of fear to critics of non-normative sexualities. Masculine subject positions in particular, due to their privileged place in society, are seen as deviant and dangerous when practised outside of the accepted societal frames. With access to power within society, there is the potential for subverting the established conditions of heterosexual males. However, by reinforcing the sexual and relationship dyad, such as through the acceptance of homosexual unions, the hegemonic conditions are maintained. Hyper-masculinity, in its display of promiscuity, power, pornography and phallocentrism, can be seen as dangerous to the power of heterosexual males through the dissolution of the feminine (and female) subject position. The problem with hyper-masculine gay sexuality may be the difficulty in assigning a female subject position; if two men are seen as equally masculine, or if masculine men are seeking other masculine men, then there is a problem in delegating a female, or feminine, position. The relative absence of the feminine on Zeus, or a desire for the feminine as object, renders all instances of male and masculine as object positions and therefore desirable and capable of being dominated by the hyper-masculine subject position. The symbolism of masculinity itself becomes a currency for a masculine sexuality that threatens the male (and supposedly masculine) heterosexual as object. Through projections of penis size, muscularity and physical and sexual positioning (who's on top as dominant), the gay hyper-masculine subject contests the very notions of masculinity as attached, and defined by, heterosexuality. Moreover, the moral panics associated with the sexual subculture as found on Zeus can be seen as fears about the extension of masculinity based on a 'conviction that audiences respond to pornographic stimuli mimetically, by enacting in real life what they see on film', thus reinscribing fears about heterosexual porn, as applied to the positioning of women (and potential violence to women), onto the practices of (masculine) men (Dean 2009, p. 114). Furthermore, there is the re-positioning of gay sex as dangerous and harmful, reinforcing the attitudes of the 1980s and 1990s with reference to HIV.

In response to these fears, and as a way of legitimising gay culture, there is a strong movement among gay conservatives to homogenise gay culture and, in so doing, establish the gay community as conformist. The idea of sexually exclusive partnerships has been sustained in the transformation of sexual liberties within society as applied to marginal sexual communities, to the point where homosexuality has become to some degree legitimised within the boundaries of homonormative behaviour and practice as they mirror heterosexual behaviour and practice. Cover explains that the shift in sexual legitimacy is no longer about the dichotomy of male/female gendered practice or hetero/homosexual identity, 'to those which contrast stable couple-like relationships with "promiscuous" sexual liaisons' (2010, p. 252). This dichotomy produces a dyad of legitimate/illegitimate identity predicated on the policing of desire. This legitimising force is borne out through research which suggests that coupling and monogamy have pervaded gay culture in Western societies, with a reported increase in gay youths desiring long-term monogamous relationships, as well as a down-swing in the reported number of sexual partners outside of established relationships (Gotta et al. 2011, p. 371). This can certainly be seen as the more publicly acceptable face of gay culture in which the heteronormative standard is closely followed, and it is suggested here that such reports closely follow the gay conservative movement with regard to the gay marriage debates.

Concurrently, however, and acknowledging the populist movement towards gay conservatism, there is a thriving culture of subversive practices and behaviours, such as clubbing, bath-houses, barebacking, circuit and other themed parties. Such subversive events and spaces have persisted historically despite various attempts at policing desire, and have perhaps flourished as agents of antagonism to the endeavours of legitimising the entire category of gay and/or queer. And this is the space which Zeus and its occupants inhabit. It is a space that acknowledges and reaffirms many of the populist tropes of desire but positions them at the edge of their acceptability. There is a play on desire here in terms of projected desire in relation to masculine embodiment, yet a reaffirmation of what is expected of masculinity and male success. This is perhaps typical of the internet whereby corporeal desires are extended through a re-embodied self of potential detachment from the everyday, yet are often reattached in the commonplace signifiers of gender, sexuality, race and class.

Conclusion

The moral panics that surround gay male sexuality and the internet are variously articulated through the site of Zeus. The displays of sexuality offered remain marginal, even as the majority culture becomes tolerant of conservative homosexuality. What participation on Zeus represents is a countervailing narrative anathema to conservative standards in terms of sexual desire. It is illustrative of a queer subculture that continues to exist outside of the tolerance and acceptance given in mass media depictions and political discourse in Western society. Whilst such acceptance should be applauded in many instances, it must also be scrutinised for its ability to reframe values, mores and ways of being along an arbitrary demarcation of what is acceptable and what is deviant. In the push for equality, especially the recent movement for marriage equality, there is a risk of homogenising queer individuality, lives and partnerships that can produce a new line of discrimination according to coupledom, gender conformity and conservative behaviours. As a result, those who prefer to negotiate their lives outside of such normative principles risk being further demonised and their practices delegitimised as deviant, unacceptable and dangerous to greater society. This occurs not only in sexual practice, however, and it is this contradiction that the site of Zeus also highlights through the absence of gender difference. Gender conformity is as much an issue of conservative politics as of sexual choice, and the absence of non-conformist avatars within the space suggests an adherence to a hyper-masculine ideal of sexual prowess. In associating the masculine ideal with sexual promiscuity, the space reaffirms the conditions of gender as established through patriarchal and heterosexist authority, as well as reiterating masculine power and agency.

5
The Female Body in Virtual Space

Introduction

In contrast to the idea that male-designed and populated spaces reinforce patriarchal conditions is the essentialist assumption that spaces designed for and by females should have different spatial goals and outcomes, as well as being utilised in a contrasting manner. This is not always the case, however, due to the implicit foundations of viewing and participatory practices that are inscribed onto space and representation. Eden: The Seduction – a site explicitly for 'lesbian and bi girls' – illustrates the conditioned conformity to standardised depictions of femininity, sexuality and the gaze. Although the space purports to be for lesbian and other non-normative female representation, identity and community (Greek Gold Lesbian Resort 2011a), the site actually promotes passive conformity to hetero-patriarchal conventions. For this reason, Eden is a space that cannot be categorised as queer, despite its intentions. Because of the continued marginalisation of non-normatively gendered and sexually diverse individuals, the categories 'queer' and 'queerness' are framed as having a political purpose, and are not simply a description of alternate sexual practice. Despite the supposed absence of males within Eden, the male gaze becomes reinscribed through the normativity of feminine expression as well as the commodification and sexualisation of the female body. The gaze is reified through the screen, as a pleasure in looking and as part of the narrative gaze as described in the formative work by Laura Mulvey (1975). Mulvey's focus is on psychoanalysis and cinematic structure but her point that '(i)n their traditional exhibitionist role women are simultaneously looked at and displayed, with their appearance coded for strong visual and erotic impact so that they can be said to connote to-be-looked-at-ness' (1975, p. 62) is highly relevant to the screen space of Eden. This chapter adds to the previous analyses of representation and space by investigating the dominant tropes of femininity within popular culture. The focus on the screen subject in relation to feminist and post-feminist discourses underpins the politics of gender that are enacted through the screen.

Specific spaces in Eden (the dance-floor, castle and shopping area) produce implied and accepted voyeurism and offer opportunities for exhibitionism. Although some sexualised conduct would be permitted given the mature rating of the space, the congruence with heteronormative, voyeuristic and pornographic representation is problematic for a supposedly queer space. The effect of such representation is to reinforce normative depictions of femininity and sexuality aligned with the male gaze, and to further marginalise female expression outside the rigid gender binary. This acts as a conservative force on potential subversive sexualities whereby non-normative identities are subsumed by the binary gender model.

Space & feminism

Rather than departing from patriarchal constraints, Eden is a space that reinforces many of the normative representations of the female subject. Framed as a female-only space, it could be expected that Eden would depart from patriarchal depictions. However, the way the environment has been created, designed and managed is similar to corporeal adult leisure space. According to Pritchard and Morgan, the male gaze is privileged in leisure space. This has an effect on bodies within leisure spaces, because 'the ways in which landscapes and destinations are imaged have significant implications for how those places and their peoples are perceived' (Pritchard & Morgan 2000, p. 886). Extrapolating the male/female binary onto that of leisure/domesticity, it is apparent that CL leisure spaces are historically sites of male privilege. Pritchard and Morgan contend that the lack of research into gender in leisure and tourist space has the effect that the male gaze and its associated privileges becomes inscribed into leisure space and therefore become the norm. The entrenchment of this norm could perhaps explain the adherence of Eden to heteronormative practices: through the absence of an alternative, the 'norm' is re-presented. There are four key areas on Eden that all work together to inscribe this notion: the beach 'as a place of adventure and sexual possibility' (Rojek 1993, p. 189) signals an implicit approval of the naked, or near-naked, form of the body; the shops, where one can purchase various clothing for the avatar epitomises the consumerist ideal, reinforcing gendered bodily representation and an inscribed objectification; the dance-club further reinforces this ideal through themed nights and events such as 'Naughty N Nice', 'Desperados' and 'Vampire Night' that underpin the site's hedonistic and sexual appeal; and the castle, which is the only structure on Eden that has secluded, private spaces, giving the assumption of a more intimate and sexual space than the other locations. Inside the castle is a smaller dance-floor with dance-poles, as well as private rooms with animations that engage avatars in sexual activity. As discussed in Chapter 1, the spatial configuration of SL, and therefore the islands and regions it contains, are inherently patriarchal constructs. An extension of techno-domination, their spatialities are assumed male and masculine. They are both techno-spaces and leisure spaces which assume the male gaze, and where pleasure is derived through mastery of the environment. The spatial constructs of Eden reinforce the foundations of spatial construction of the corporeal; Eden is an extension of the pleasure periphery (Figure 5.1).

This continuation of normative spatial representation extends to avatar representation where the appearance and performance of avatars on Eden also take a heteronormative function: that of the feminine. The

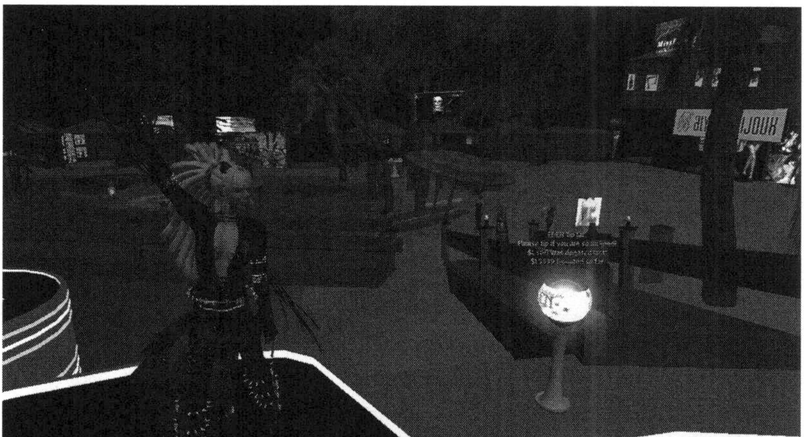

Figure 5.1 Dance-floor, Eden
Source: With permission from Linden Lab and the residents of Second Life.

absence of androgyny or gendered difference on Eden is a questionable phenomenon in a self-defined non-normative space, particularly when taken alongside the other heteronormative aspects of the environment. If you were unaware upon arrival on Eden that it was a female-only and minority space, you could be forgiven for mistaking it for a heterosexual space. The male gaze is inscribed in the environment and in the appearance and performance of the avatars. Although it is possible, as part of a postmodern and subjective account of the gaze, for a female to embody a masculine subject position in her looking, the saturation and ubiquity of the feminine form detract from any sense of a variety of choice in her own performance, thereby reinscribing gendered positions. This construction of gender is indicative of the heteronormative practice of keeping femininity and masculinity apart, diluting the potential for queer expression as multitudinous. It is ultimately this absence of choice, limiting identity expression, that is so necessary in spaces designed and utilised by those of subjugated groups. Although Eden is a space restricted to female avatars (the possibility for CL male infiltration of the space is discussed later), the space is populated for the most part by hyper-feminine female avatars. The customary appearance of avatars on Eden is sexualised. They wear few clothes, revealing much of their skin, and emphasise curvaceous and youthful figures. In addition, most avatars have long hair and reflect a CL heteronormativity.

There is an incongruence between Eden's self-classification as a space for 'lesbian and bi girls' and its usefulness as a space of tolerance and acceptance for a spectrum of female-identifying individuals. Whilst Eden is a space for certain types of females, particularly those identifying with modes of sexuality divergent from the norm, it cannot be categorised as queer. As elaborated by Jodie Taylor (2008, p. 653), defining queer is important as it identifies the embedded ideological foundations of identity:

> Both Jagose (1996) and Phelan (1997) recognise that queer is particularly concerned with interrogating the notions of fixed, coherent, unified and transparent identity categories. Butler (1990) however, reminds us that gender and sexual identities are in no way natural or stable, suggesting instead that gender is constructed by a series of repeated gestures understood as performative. Queer is not lesbian and gay, suggests Butler (2001), but rather is an argument against lesbian and gay specificity and the idea that being lesbian or gay presupposes that we perform our gender and sexuality in a particular, fixed manner. There is, of course, contention over the usefulness of these

> definitions, because some scholars have argued that such ambiguous theoretical suppositions fail to account for lived experience.

Perhaps in a perfect world identity categories would be made obsolete, but Butler's contention does lead us to suspend a strict adherence to categories that re-normalise subjugated groups within new boundaries of adherence. Whilst acknowledging that an all-encompassing dissolution of gender and sexual categories is not possible, at least at this moment in time, queer as a descriptor can be used as a destabiliser of categorisation, and so can be understood as a marker of fluidity in an environment where a spectrum of gender and sexual positions is represented. This has practical outcomes for those identities that cannot, and will not, be subsumed into homogeneity, even at a sub-classification level (such as by various identifiers and markers of queer, lesbian, bi- or gay performance).

Spaces such as Eden initially seem a welcome environment as they potentially provide a unique destination for a global queer female community. Lesbian spaces in CL are often less visible and less commercial than gay male spaces and heterosexual social spaces, which can produce a difficulty in accessing community for those not familiar with 'the scene' (Valentine & Skelton 2003). In contrast, spaces such as Eden on the SL platform are relatively easy to access, with the benefit of anonymity. Tom Boellstorff (2008, p. 165) suggests that SL may provide a 'virtual closet' where people 'could live out same-gender desires they were unwilling or unable to enact' in CL. He further adds that 'Second Life could act as a venue to grow comfortable' with an alternative identity 'before coming out in the actual world' (2008, p. 165). As a site of initial engagement with the (female-identified) queer community, however, spaces such as Eden are unrepresentative due to the feminisation and sexualisation of female representation. This may be counter-productive in the 'coming out' process of youth, or even in the continuing engagement with community for those living outside of cities that often foster visible gay/queer communities. The uncertainty of queer youth in 'coming out' is explained by Valentine and Skelton (2003, p. 852):

> First, the majority of young people are born into, and grow up within, heterosexual families where the expectation is also that they too will be heterosexual. As such they rarely have any direct contact with lesbian and gay men and therefore have little knowledge or experience of alternative sexualities and what it means to live a lesbian and gay lifestyle.

The problem with Eden as a space of community and 'coming out' is that it is replete with heteronormative and hyper-gendered forms inconsistent with the subversion of gender found in many queer CL spaces. 'The scene' as characterised in the queer community in its bars, pubs and social spaces 'not only provides a transitional space where young lesbians and gay men can express their self-identities, but also offers a space where others can validate these identities' (Valentine & Skelton 2003, p. 854). In taking SL queer spaces, such as Eden, as an extension of this scene as a place of social engagement, then it can certainly be problematic for identity formation and internal cohesion, particularly for androgynous individuals who may feel ostracised from the space due to its emphasis on the body in terms of hyper-gendered forms.

As noted by Valentine (1996) and Casey (2004), there is a need for queer female space as separate from other spaces, even those designated as gay. Often lesbians and bisexual females utilise mixed gay and queer spaces due to the limited number of women-only queer venues and/or events that exist which, according to Casey (2004), can sometimes create tension between queer males and females. This, alongside the recent phenomena of the de-gaying of gay ghettoes by heterosexuals seeking spectacle, has made queer female areas difficult to maintain on any regular basis. For spaces set up for minority cultures, 'infiltration' can effectively impact on that culture and, in the case of SL where anonymity prevails, it can have the effect of eroding trust in the space itself and those who use it. It has been proposed, as part of SL's age verification debate, by an owner of a female-only space that SL should introduce voluntary gender verification to prevent lesbian spaces being infiltrated by CL males. However, this is not current policy in-world, so the possibility of males embodying female avatars is on-going (Fine 2010). Although there is anecdotal speculation on SL blog sites that you can tell a male-in-drag by their avatar's breast size (with the assumption being that large-breasted female avatars are more often men), there is no real means of gender identification. In addition, there is the very real possibility that males-in-drag frequent female-only spaces for sexual purposes which can be very unsettling for queer females using the space for community or interaction. As described by the aforementioned age verification forum, this kind of masquerade can have emotional consequences, such as the feeling of cyber-rape.

There is an assumption that the creation of lesbian spaces in the virtual realm would be outside of the same constraints imposed on communities of difference in CL, where real-world boundaries, judgement and intrusion of heteronormativity impose themselves onto such spaces due to geography, the small amount of gay- and queer-dedicated spaces (especially female-only spaces), and the entry of heterosexuals into these spaces. Virtual spaces, by contrast, potentially minimise these constraints and intrusions through removing the geographical boundaries of travel and entry into the space. For instance, one does not have to travel through the potentially threatening streets of the city in order to access the queer space, or gay ghettos, of CL – research by Tomsen and Markwell (2009, p. 202) 'suggests prejudice-driven violence directed against gay men, lesbians and other non-heteronormative groups is widespread' despite the growth in representations of such identities in the mass media. In addition, due to Eden being a female-only space, there is an initial expectation that only bisexual and lesbian females will use the space. Additionally it is assumed that avatar appearance and performance will reflect queer sensibility in CL. Given that SL allows for the expression of appearance outside that of CL physicality, there is the potential for displays of androgyny and subversive modes of gender more radical than those seen in the corporeal environment, displays that could be read as even more queer than those found in environments surrounded by conservatism. Valentine and Skelton remark on the importance of queer performance on the scene (2003, p. 855):

> The scene offers young lesbians and gay men an opportunity to step out of the hetero-normative world where they often feel marginalized. Clubs and bars provide spaces where people can lose themselves and their troubles in music, dance and sex. They are expressive, performative spaces where people can enjoy themselves together in ways that can be empowering.

As a site of queer validation, Eden is problematic. Although it is necessary to acknowledge all forms of expression of individual queer (or not queer) expression, the minimal performance of a queer sensibility often seen through androgyny and subversion, prompts curiosity regarding the usefulness of the space as a site of community, solidarity and transgression. As a result of this critique, it is perhaps more useful to categorise Eden less as a lesbian or queer space, and more as a space of spectacle. If queer space is defined as a space where one can feel comfortable in one's individual expression of sexuality outside of normative identity performance and sexuality, and lesbian space as a site where one can express a female-centred queer identity, then Eden falls outside of these parameters. Instead, it is offered that Eden is a space for heteronormative female-to-female sexual practices.

Representation of the (queer) female subject

Despite being a space only for female avatars, female expression on Eden tends to be hyper-sexual and feminine. Eden is thus analogous to many mainstream representations of a fetishised female-to-female sexuality, and ultimately undermines the performance of alternative embodiment and ways of being whilst reinforcing prevailing patriarchal assumptions about female appearance, performance and sexuality. This is illuminated in those issues of the space that can erode the foundations of the site as a useful, safe and tolerant space for all lesbian and bisexual females. Such issues are related to the representation of the feminine norm, and the internalisation of the gaze, which include: the limiting of a female queer identity to a feminised and/or heteronormative expression; the reduction of queer sexuality to that of novelty; tensions between lesbian and bisexual identity; the 'infiltration' of queer space by both heterosexual females and CL males embodying female avatars; and the reading of sexuality of the site as heterosexual pornography for the male gaze:

> Men look at women. Women watch themselves being looked at. This determines not only most relations between men and women but also the relation of women to themselves. The surveyor of woman in herself is male: the surveyed female. Thus she turns herself into an object – and most particularly an object of vision: a sight. (Berger 1972, p. 47)

For these reasons, Eden may be problematic for queer females in terms of identity and expression, and instead of being a space of sexual freedom may become an extension of their sexual identity as practised in CL spaces. There are alternate viewing positions, articulated in critiques of Mulvey's and Berger's work, that position the female as powerful in a position where she looks upon others and holds power in being looked upon, particularly as a non-heterosexual subject. (Evans & Gamman 1995) However, it is the contention here that such a position is ultimately marginalised by the dominance of the fetishized spectacle of the erotic female.

As described by Barbara Creed, the lesbian body in popular entertainment is 'variously depicted as narcissist, sex-fiend, creature, tomboy, vampire, man-eater, child, nun, virgin'. She also asserts that 'one does not need a specific kind of body to become – or to be seen as – a lesbian' (Creed 1995, p. 87), meaning all females have the potential (or threat) of lesbianism. This pronounces a threat to the social order of heterosexual marriage and male sexual dominance, whilst at the same time presenting a disruptive sexual deviancy to which males desire access. If access is denied, lesbianism fulfils the promise of existing outside of male desire. Feminised depictions of lesbianism, however, are seen as less threatening and symptomatic of what Creed defines as the 'narcissistic lesbian body' where 'the other woman symbolise(s) her own reflection' (1995, p. 100). The threat of the narcissistic body, unlike the masculine lesbian body, is overturned through depictions of femininity, and instead becomes ambiguous and desirable (Creed 1995, p. 100):

> The narcissistic femme lesbian, however, almost always adopts an ambiguous position in relation to the gaze of the camera/spectator. She is on display, her pose actively designed to lure the gaze; the crucial difference is, however, that the spectator is shut out from her world. He may look but not enter. Images of the lesbian double are designed to appeal to the voyeuristic desires of the male spectator.

More recent depictions of the 'narcissist female lesbian' in popular culture have become the norm for lesbian representation. Lisa Diamond argues that the pervasiveness of hetero-flexibility 'can have the effect of trivializing and depoliticizing same sex sexuality by portraying it as a fashionable 'add on' to otherwise conventional heterosexuality', and that 'such images implicitly convey that the most desirable and acceptable form of female-to-female sexuality is that which pleases *and* plays to the heterosexual male gaze' (2005, p. 105). As made explicit by Diamond, such representations of femininity within queer boundaries can have a positive effect in showing the diversity of the queer female body. However, this becomes problematic when it is the only depiction of female queerness and otherness, as it reinforces heteronormativity and desire that is inextricably linked to the male-gaze. Furthermore, 'it presents girl-on-girl action as exciting, fun, but, crucially, as entirely unthreatening to heterosexuality' (Evans & Gamman 1995, p. 153).

For non-heterosexual women, spaces that represent heteronormativity reinforce standards of appearance and behaviour experienced in most facets of their lives. As noted by Valentine (1993), this is dealt with by having multiple sexual identities in order to traverse the complexities and difficulties that can occur due to being 'out' in various places. Levels of comfort in performing identity range across the public spaces of work and leisure and the private space of home. Valentine describes the performance of sexuality over both time and space, where the individual monitors their appearance as well as their behaviour according to the suitability and acceptance of queerness and/or otherness of the environment that they are in. In considering the site of Eden as typifying heterosexual gender expression, it may be necessary for some queer females to play up to this representation, and thus extend their CL performance of sexual and gender fluidity. This produces a contradiction of form and function of queerness: avatars are expected to fit the feminised model of appearance but are also expected to perform queer sexuality. Furthermore, there is an alignment with the power relations entrenched in CL public space highlighting 'how sexual identity is strongly linked to gender identity and hence heterosexuality to patriarchy' (Valentine 1993, p. 246). For such a significant number of avatars to assume a normative feminised appearance suggests that the power relations of CL spaces not only superimpose themselves onto virtual sexualised space, but can even be amplified.

One of the key problems with Eden as a queer space is in the normalisation of hyper-feminised avatars and complicity to gender norms, whereby the space-value of queer space can be categorised through the potential for transgression, embodiment, identity and subversion outside of the established norms of patriarchal sexuality. Instead, the site of Eden frames queer sexuality as a novelty, as a supplement to heterosexual desire rather than as a replacement; female-to-female sexuality is seen as exotic. The normative gaze of the heterosexual male onlooker is where the performance draws upon notions of desire. Whether a male is present in form is irrelevant; the gaze is omnipresent and normative, allowing for what is acceptable and what is not. The power of avatar performance comes from an adherence to patriarchal desire, in that 'the sexualisation of lesbian bodies ... seems to be constructed in relation to heterosexuality not as an autonomous or independent sexual identity' (Evans & Gamman 1995, p. 153). This site of representation reinforces Rosalind Gill's description of lesbian images in advertising where the dominant representation is that of the 'hot' or 'luscious lesbian' (2008, p. 151). This figure 'is notable for her extraordinary attractive, conventionally feminine appearance. Women depicted in this way are almost always slim yet curvaceous, flawlessly made up and beautiful' (2008, p. 153). This commodification of female-to-female sexuality is an extension of the framing of heterosexist desire and eroticism that bases the pleasure at the site of the gaze and the objectification of the feminine body. It appears as an extension of the controlling and surveying mechanisms of the visual; a replication in form of the 'right' way to display the erotic female body.

Sexualisation, voyeurism and commodification: The implied male-gaze

The implied gaze is directly linked to voyeurism. It can be argued that the visual nature of the internet and 3D visual environments implicitly encourage passive viewing behaviours as well as the ability to explore taboos and fetishes. The relative anonymity of the internet emboldens users to engage in behaviours that they may have previously been reluctant to entertain. Unlike stalking or pathological voyeurism, however, SL users participate knowing the potential of the gaze and often display reciprocal voyeuristic behaviour. In Eden, voyeurism is enacted through the patriarchal gaze and female desire as subsumed within the eroticisation of female-to-female interaction. Voyeurism has become a normative activity in relation to visual culture especially as part of screen culture. However, according to Jonathan Metzl (2004, p. 419), the gender implications have remained uncontested in view of the Freudian assumption of anxiety and control:

> Critiques that presuppose these two categories (normative and pathological) fail to see, for instance, how reality television shows, twenty-four-hour Webcam internet sites, and other examples of the voyeurism popular culture has so effortlessly adopted as an autobiographical adjective often re-enact the very gender anxieties identified, and reproduced, by psychoanalysis, while blindly pathologizing threats to normative heterosexuality and heterosexual marriage.

Taken in the above context, voyeurism on Eden is largely normative through its re-construction of the girl-on-girl scenes reminiscent of mainstream heterosexual pornography. As noted earlier, Laura Mulvey asserts that women are traditionally given the exhibitionist role in viewing pleasure, their bodies coded for 'visual and erotic impact' (1975, p. 62). As such, these scenes reinscribe feminine performance to that which is defined through and by male desire.

There are three spaces on Eden that allow and reinforce voyeurism: the dance club, the castle and the shops. The club has a stage that features a DJ and often has the 'Eden Sisters' (the managers of the space) overlooking the dance-floor. Dancers utilise animations that enable their avatar to dance erotically so that the club visually represents a space of seduction, akin to the club scenes in mainstream music videos, or their 'space of action' as described by Coy (2009, p. 378) in her discussion of the sexualisation of young women:

> Girls' space for action is ... shaped by expectations and ideals that limit their life, career and self-aspirations, and the relationship with their body, to those plastered over advertising billboards, music videos and reality television, with few alternatives presented as desirable.

It is not unusual to see avatars in embrace on the middle of the dance floor, drawing attention to their sexual display and exciting similar behaviour in others. The dance club replicates masculinised space for viewing pleasure. Performance is monitored by the 'Eden Sisters' who are situated above the dance-floor, looking down in tacit approval of the 'action' below. In addition, avatars that are situated on the periphery of the dance-floor, encircle the space of performance with their collective gaze.

The castle offers an even more sexualised space. As the only enclosed space on Eden, it signifies a site of seclusion and sexual encounters too risqué for the open environment. Congruent with this suggestion is the interior of the castle itself: darkened rooms, a smaller more intimate dance-floor equipped with dance pole, erotic pictures adorning the walls, and couches with animations (pose-balls) for kissing and cuddling make up the decor of the main entrance hall. In addition, the castle has various smaller, private rooms equipped with beds or sometimes baths, as well as animations for sexual intimacy. The castle is a clichéd spatial configuration, signifying commodified sexuality. The possible sexual performances, permitted by avatar animations in the castle's spaces, reproduce the codes of female-to-female sexual interaction in hetero-normative media. Although the castle has the appearance of seclusion, this is more suggestion than fact; it is possible to view all spaces of the castle at any time. This is possible due to the open access of the spaces, and also due to the capability of the viewing toggle (part of the camera feature) of SL's interface. As a result, privacy is not possible, suggesting that voyeurism may not only be tolerated, but implicitly encouraged.

Even in the less sexualised spaces of Eden, there is a voyeuristic element that underscores all appearances and interactions. As if the emphasis on the seductive body is not explicit enough, there is a shopping precinct on Eden where one's avatar may purchase clothes and get dressed accordingly. All shops feature hyper-feminised bodies, and there is a particular emphasis on clothes and accessories (including avatar skins that give an overall shape colour and texture to the body) that accent taut, youthful bodies. Femininity is stressed through bodily contours, clothing shapes and the prominence given to displaying the breasts, legs and bottoms that so often feature in music videos, advertising and pornography (and which are often exaggerated past the scope of average female biological capacity). Such representations of the avatar reinforce the commodity value of the body, not only in CL but also on SL. Spaces like Eden can be seen as an extension of real-world desires for bodily perfection and sexuality as invested in the female body. It is unsurprising, given the foundations of SL community being based on its economy, that bodily representations reinforce those on offer in the highly commoditised industries of raunch culture (Williams 1999) that permeate mass culture in capitalist societies. The female body is commodity; it is displayed to be looked at, to be desired, and to be accessed for pleasure.

The emphasis on looking and deriving pleasure from the female body on Eden is a relatively short extension, given its interactive potential, from the voyeuristic practices of mass visual culture. Jonathan Metzl contends that such voyeurism is 'a practice that is culturally pathological, imbued with power, gender and other types of nonchemical imbalances that let us see the voyeur as an exaggerated extension of society as well as an aberration from it' (2004, p. 428). In a shift from pathologising all forms of voyeurism, it is now culturally, as well as medically, accepted that some types of voyeurism are acceptable, if not inevitable in such a society. What Metzl (2004, p. 430) highlights is the 'pathology' of our culture in its imbalances and inequality:

> Always watching all of the time, popular culture enforces the seeming silence of looking, and the hegemony of a gender ideology that (under the guise of all-access viewing) tirelessly recuperates an often-Oedipal marriage myth and marks other forms of temptation as deviance.

To this point in the discussion, the male-gaze and patriarchal power have been assumed invisible in the representations of the avatars as well as the spatial and discursive configuration of Eden. The shop display below (Figure 5.2) is typical of the type of sexual products available for female avatars. Taken as a ubiquitous force in the organising principles of space and identity, hetero-patriarchy is assumed to function even in the absence of male embodiment and manifestation. However, as a central element of voyeurism as well as the fluidity of identity as applied to the virtual realm, the idea of male presence must be considered.

In addition to the implied male-gaze in reference to discursive spatial structure and modes of representation, there is the possibility of infiltration of Eden by CL males embodying a female avatar. This can have a profound negative impact on culture in spaces set up for the use of marginalised identities. In the case of SL, where anonymity prevails, it can limit trust in the space and those who use it. The effects of such infiltration and gender-swapping can have real psychological effects on users, such as the case reported by Mandy Dorfler (2010) who married an individual in SL whom she thought was a female:

> Too often women are violated by men in SL. Men come to women only areas and use a female avatar to cause troubl [sic] emotionally and in other ways. This is a form of cyber rape ... I married what a thought was a German girl. Spent a fortune on the wedding and other things then 'she' confessed to being a rl male.

Although it was proposed by Dorfler, and others, that SL should introduce voluntary gender verification to limit lesbian spaces being infiltrated by CL males, this is not current policy in-world, so the possibility of males embodying female avatars is an on-going issue. Dorfler states that such deception happens 'often' in SL according to anecdotal accounts; however, it is difficult to verify this with any certainty. What remains is a very real potential for CL males to frequent female-only spaces for sexual, or other, purposes which can have very unsettling consequences for lesbian/queer females using such spaces for community or identity expression.

Specific research into SL identities by Ducheneaut et al., has found that 'gender-swapping is particularly prevalent in SL, with a large number of male players favoring female avatars' (2009, p. 1152). In the case of an environment based on sex, however, the reasons for gender swapping may be based on the sexual tastes and preferences of the individual. It may be that heterosexual males wish to use female-only sites as a way of viewing (and participating in) female-to-female sexual interactions. Infiltration may explain the phenomenon of hyper-gendered avatars in an environment set up for queer sexualities. The enduring possibility of such infiltration can create a sense of unease and untrustworthiness in SL environments, with the added presence of heteronormative avatars, increasing the potential sense of un-belonging an individual may feel

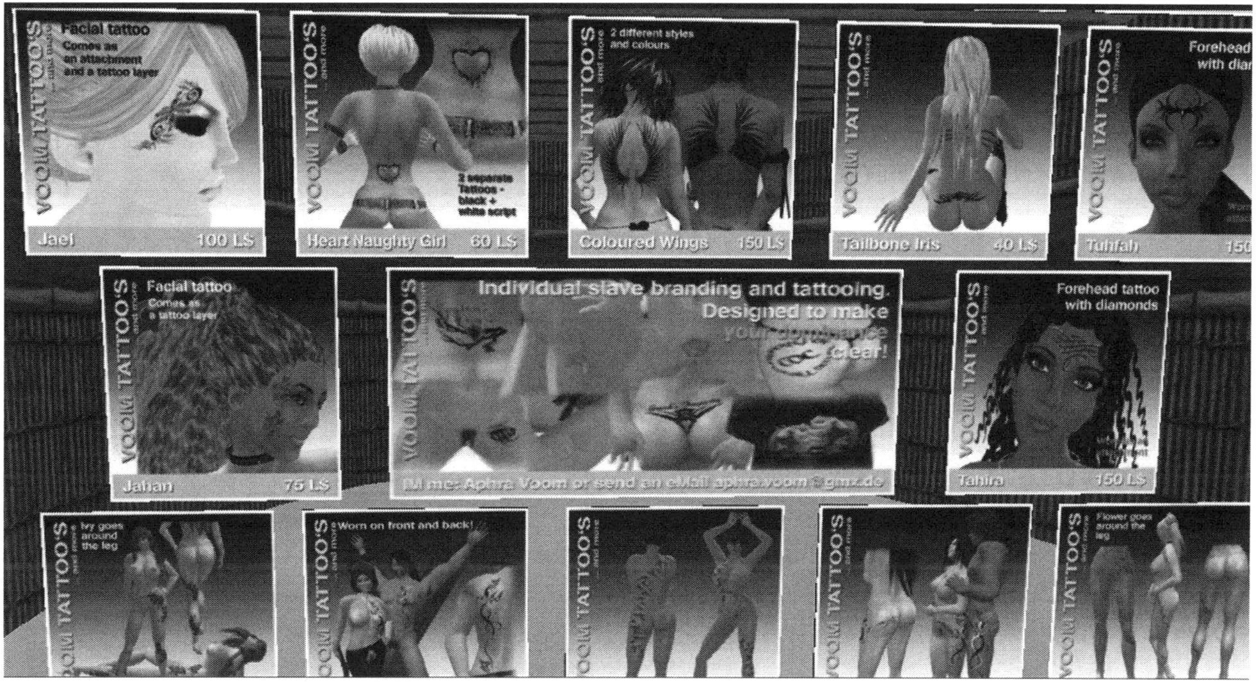

Figure 5.2 Advertising board, Eden
Source: With permission from Linden Lab and the residents of Second Life.

when visiting a supposedly queer space. In describing straight women in the CL gay space of Manchester's 'Village', Binnie and Skeggs note that although 'the village constitutes a comfortable space for some of these (straight) women, their sense of comfort and ease in the village may produce discomfort and disidentifications among other users of this space' (2004, p. 54). If this is the case for a space such as Eden, the result may be that the space is utilised less by queer individuals, including those looking to experiment with their sexuality beyond that bounded by patriarchal determination, and more by CL heterosexual males utilising the space as an extension of mainstream pornographic practice.

Pornographic representation

Due to the (almost entirely) visual nature of sexual interactions, Eden and SL in general have obvious parallels with pornography, albeit in a more interactive form; it is possible to view and/or participate in sexual activity rather than being just a passive consumer of images. In analysing Eden through the lens of mainstream pornographic discourse, the aspects of sexual economy and consumerism alongside sexual agency reveal the complexities of both lesbian and female visual performance in the SL environment. The difficulty in Eden achieving value as a queer space comes into question through the hetero-normative display and performance of avatars that appear contradictory to queer expression: 'Ciasullo (2001) and others (e.g. Garrity 2001; Jenkins, 2005) argue that representing lesbian women within heterosexual attractiveness norms constructs them for consumption by the male gaze, in much the same way that heterosexual women may be cast as eye-candy for a male audience' (Jackson 2009, p. 201). The space replicates the porn industry's representation of female-to-female sex, including the coding of lesbianism as an aside to heterosexual sex. Various markers abound on Eden that realise this extension of the mainstream: the club where near-naked avatars dance in a highly sexually suggestive way; the shops that promote femininity and sexual exchange value centred on bodily appearance; and the castle that replicates heterosexual markers of sex, availability and titillation through the intimate spaces, secluded rooms and a performative dance space equipped with twin dance poles. The entire site reproduces the signifiers of sex and femininity as seen in mainstream pornography and film, with the only difference being that the male position is purely implied (with the exception of CL males embodying SL females).

Expressions of appearance and performance in much lesbian-centred pornography often challenge the depictions seen in mainstream discourse. This is because the material is generally created by queer females for the consumption of queer females so that the subsequent depictions of gender and sexuality lie across a spectrum of practice and are not restricted to portrayals often found in heterosexual porn that privilege the male-gaze. Carol Queen and Lyn Comella describe the interest for sex-positive females to not only fight against conservatism, but to renegotiate the power structures or pornography, suggesting that female-generated porn 'was also an attempt, on a really feminist level, to take the means of production and put it into the hands of women' (Queen & Comella 2008, p. 285). This has the effect of altering gender and sexuality expression by transgressing male-to-female binaries, and queering it through offering expressions of butch, femme and androgynous representation (as well as BDSM and use of toys). In depicting sexual identities and practices outside the norm, lesbian pornography inverts many of the symbols used in hetero-normative porn such as the rigid gender binaries that are fused with depictions of power, penetration and the phallus. Although Corsianos (2007, p. 868) argues that lesbian pornography reinforces patriarchy through the replication of certain scenes, she does not consider the transposition of pleasure associated with these scenes. Sites of pleasure are incongruent with normative portrayals of dominance as typically aligned with gender: the butch/femme dynamic can often produce more scenes of pleasure for the femme, whilst her more masculine partner, although in the gendered position of control and power and thus assumed pleasure in the hetero-normative function, is positioned to be the provider of pleasure rather than the recipient. Meanings become ambiguous through the disruption of signs, and although much queer porn replicates scenes from straight porn, this ambiguity unsettles and reforms the contingent messages and agency attached to the normative patriarchal function of porn in the mainstream.

Where this becomes problematic is in the production and consumption of values associated with the visual representations of the pornographic body, and whether these representations either contest or reinforce contingent power structures. Therefore, the question that is invariably posed concerns whether a site of heteronormative representation can be a valuable space for the expression of queerness and the inversion of power and patriarchy that queerness can embody. According to Linda Williams, meaning within pornography and the potential for 'real' sexual expression can be attributed to the non-commoditisation of the form. In describing video pornography, she writes:

> Indeed, all forms of pornography tending toward the amateur end of the spectrum are likely to deliver 'something' because of their valuation of authenticity over glamour (no breast jobs allowed). These videos tend toward the *unique* and *local*. (2007, p. 76)

She maintains that non-conformity to established pornographic conventions of hyper-real simulated sex acts, whether straight, gay or any type of subversive sex, hold intrinsic value to culture and representation. Such a contention is problematic to spaces such as SL: representation is entirely hyper-real, and all sex acts are simulated. This can lead to the assumption that all SL sexual activity is inauthentic, as a consumption of 'nothing'. However, due to the interactive ability of individuals to enact sexually, and in real-time, with their avatars, this is too limited a description. For those who invest temporally, emotionally and financially into their avatars and SL experience, this position is too reductive to explain their sexual experiences. The point of contestation, then, rests upon the lack of difference, especially for sites like Eden which are explicitly set up for queer experience. It is not the medium that is problematic, but the message in its standardised representation.

For sex to be truly queer, in the sense that queer is a contestation of hetero-patriarchal conventions, it needs to destabilise the established power relations that are its referent and not simply to fetishise them. Carol Queen and Lyn Comella argue that pornography should aim to replicate the values of the community it is produced for. They recount the process for selecting videos for sex store *Good Vibrations* (Queen & Comella 2008, p. 289):

> The other thing that we really had to grapple with, and this did have some effect on us, was when the porn industry, in either its alternative or its more mainstream guises, creates something around a sexuality that is already a little marginalized, but it doesn't feel respectful, like tranny porn or grandma porn or fat girl porn being fetishes; almost none of it feels like it carries the values, or the value, of that community into the sex you're seeing on screen.

The absence of difference, and the ultimate fetishisation of lesbian sex, renders the space of Eden more akin to mainstream pornography in its normative sexual expression than as a queer space which privileges subversion. The queering of space requires a challenge to compulsory heterosexuality as well as gender normativity and the privilege of the male-gaze. Whilst Eden may initially seem to present a challenge to compulsory heterosexuality, it conforms too easily and readily to normative depictions of both gender and mainstream visual sexual performance. In addition, it does not provide agency for sexual and identity expression outside of the social and cultural models provided in consumerist society. This mainstream aesthetic is described by Marilyn Corsianos in her work questioning women's sexual agency in reference to pornography (2007, p. 865):

> Most of the scripts are written by straight males catering to a predominantly straight male audience. And, therefore, it is no surprise that a particular female aesthetics is promoted in mainstream pornography. For instance, the female actors often have long hair, are thin, often Caucasian, 'young' (usually between teens and 30's), have breast implants, wear lingerie, high heels and plenty of make-up. Their legs are shaven and their vaginal area is usually partly shaven with the remaining pubic hair trimmed short.

Although describing mainstream heterosexual pornography, this description could well describe the appearance of many avatars on Eden. The effect that this has is to reduce the sexual agency of the individuals who use the space, minimising diversity and difference and producing a conformist conception of gender and identity. This is particularly the case due to the lack, and sometimes absence altogether, of androgynous or masculine female avatars. More than any other gender category, the masculine female destabilises gender and sexuality by subverting the power associated with masculinity.

The absence of masculine bodies on Eden can be linked to the greater societal disapproval for females, including lesbians, assuming power and therefore qualities associated with masculinity. This can be attributed to the great social contention that if a female performs masculinity she wishes, in fact, to *be* a man. The Freudian conceptions of inversion and penis-envy pathologise expressions of androgyny and butch performance to the extent that it is a subject position feared more than effeminate men, as displayed in the great absences of masculine female performance in mass culture. Whilst the uber-gay, effeminate male, although often humourised, has gone mainstream in TV shows like *Modern Family* (2009–) and *Glee* (2009–), there are minimal representations of the masculine female. According to Judith Halberstam, this is because the masculine 'has been reserved for people with male bodies and has been actively denied to people with female bodies' (1998, p. 269) and is attributable to the retention and reservation of power. In returning to pornographic representation, it may be that this phenomenon (a reluctance or aversion to appropriating the powerful position) explains the absence of masculinised avatars on Eden, whereby non-feminised appearances and behaviours are assumed to be the reserve of men, and to hold patriarchal significance; it may be that a more feminised space frames all practices within that space as being outside of the problems of power and the gaze. Paradoxically, however, it is the very absence of non-feminised forms and practices that upholds this very patriarchal construction, and in turn reinforces its privilege. Contrary to Corsianos's contention that sex acts between females which visually simulate masculine and feminine roles reinforce patriarchy and therefore limit sexual agency, it is this very transposition of gender performance that makes the pornographic representations queer and therefore subversive to hegemonic power; the inversion creates a destabilising of the function of gender and sexuality.

Eden, however, maintains the hetero-normative model of performance for female sexual expression; there is no inversion of gendered agency (in terms of Freud's pathological inversion) and no destabilisation of the masculine/feminine dichotomy. The space is questionable as queer due to the homogeneity of form. It may even be seen as intimidating for those looking to seek out a space for community of difference that is a departure from their CL experience. What Eden does provide, however, is an extension of the commodity-value of sex and sexuality that may be used by either males (in drag) or females, straight or otherwise, who seek an eroticised sexual experience. It is unsurprising that the economic basis of SL produces replicates in form of mass consumerist ideals, because it is essentially an extension onto the screen of the hyper-real portrayals of sex and the body in other media forms and in the pornographic industry (Corsianos 2007, p. 864):

> We do not all share the economic and ideological power needed to promote particular sexual performances, body images and definitions of sexuality, and to profit from them. It is evident that a particular gender and sexuality hierarchy continues to exist on various levels by the mainstream pornography industry estimated to be an 8 billion dollar a year industry (Juffer 1998) to as much as 13 billion (Cornell 2000). For instance, mainstream straight sex and mainstream lesbian sex are created and promoted in very specific ways

by the pornography industry, instilling in the minds of many what 'women' look like and how they perform sexually.

As part of the capitalist and commoditised nature of SL, Eden plays into consumerist sex models of mainstream mass representation that can be bought into by any gender or sexuality. Although the acceptance of sexual difference has come a long way, this acceptance has conditions and remains associated with bodily perfections associated with heterosexist ideals of female beauty. The idea of perfection is closely linked with the name of the space, and is something that needs to be acknowledged in the reinscription of patriarchal fantasy.

Eden as patriarchal fantasy

The name of Eden, in association with the dominant modes of representation and spatiality of the site, signifies myths of gender and sexuality in Western, Judaeo-Christian discourse. These myths are centred on the notions of both pure and deviant sexuality almost exclusively attached to heterosex. Eden most obviously references the Biblical narrative whereby Woman is painted as sexual temptress who leads Man astray. In reference to the Genesis tradition, pure sexuality is made deviant through the enticement of the female. In summary of its ideological imperative, Kevin Harris (1984, p. 96) states:

> And it also appears that the generalisation from Adam to all of humanity is quite in order: one does not have to go too far between the lines to pick up the message that taking notice of their wives is not a very smart thing for men to do, and that just as Eve led Adam and the whole of humanity into sin and suffering so too do all women have the propensity to bring about sin and transgression in the world.

Using the creation myth as symbolic, SL's Eden is a play on deviance, where female avatars represent forbidden sexuality and enticement. In the 'godless' world of Western consumerism and hedonism, the female body acts as the ultimate vehicle for participation in bodily pleasures. Much of the signification is reliant on the very signifier 'Eden' as it is known through the Western Judaeo-Christian culture. Although some of the imagery of SL's Eden draws upon the Biblical symbolism in Genesis (there is a river running through the site and there are a number of trees beautiful to behold), most of the space is characterised by Western commercialised space similar to a holiday resort: the shops, dance-floor and castle are the major features, much more so than any of the 'natural' elements of gardens. This is suggestive of the transference of 'Eden' as a signifier of devout purity in the common imagination to that which is pleasurable.

Additionally, the ideas of control of men over women as detailed in Judaeo-Christian culture (most of which post-date the initial Eden narrative in Genesis) are subverted through the environment of SL's Eden. The Biblical narrative suggests that 'In the conflict situation arising from the disruption, due to sin (enacted through eating the forbidden fruit), of the relationship between the sexes, the woman will, in general, be dominated' (Evans 1983, p. 20). Moreover, due to the sin predicated by woman (that of Eve), women in general need to be controlled as a necessary condition of society. Mary Evans explains that the consequences of sin provide 'an explanation, although not necessarily a justification, for the male domination in Israelite society', and that 'Where there had been communion there would now be conflict; where there had been equality there would now be domination' (1983, p. 20). It could be taken that the progressive attitudes towards sex and intimacy in the SL environment are closer to the Genesis version of Eden before the Fall. If, as Evans suggests, women could be understood as being equal with men prior to the 'disruption' of the forbidden fruit, then SL's Eden could be considered to be returning to that state. Underlying this assumption however, is the invisible but omniscient gaze of man. Although (self-acknowledged) CL men are not permitted to visit the space, there is an ever-present performativity towards male desire. In the absence of male representation, there is an implicit maleness assumed in the female avatar performance that is predicated on heteronormative desire of female beauty, sexuality and female-to-female sexual display. So, even if SL's Eden is a return to the equality of Eden before the Fall, this quality is still premised through male desire and the normalised relations of gender.

A more recent take on 'Eden', however, appears to challenge conformist representations of gender and sexuality. Ernest Hemingway's novel *The Garden of Eden*, published posthumously in the 1980s, follows the relationship between an American writer and his wife. Honeymooning in France and Spain, their relationship diverts from the 'normal' course of marriage, as is expressed in the desire of both newlyweds towards a young Frenchwoman, ultimately ending in estrangement for the couple. Unlike the Biblical narrative which denotes strict gender and sexual binaries, J. Gerald Kennedy (1991, pp. 187–188) notes that Hemingway's tome reveals a fluidity of gender and sexuality:

> [H]e [the American novelist] plays a submissive role in intercourse as his wife Catherine assumes the name 'Peter,' mounts him, and calls him 'my girl Catherine'. He elsewhere calls her 'brother' and submits to her scheme of effacing gender differences; the two dress alike and sport identical coiffures of short hair bleached pale blond. Subsequently they take turns betraying each other with Marita, the dark-haired bisexual who loves them both.

Beginning from a place of happiness, the sexual digressions that are taken by David and Catherine throughout the novel disrupt the 'paradise' from which they have come. Although appearing as a place of Eden in the Biblical sense of nature, purity and peace, their journey has led the protagonists into deviance. There are strong similarities here, with SL's Eden and the Biblical narrative, through the ideas of non-normative sexuality and gendered practice where the 'protagonist's fall from apparent innocence into sexual experimentation and duplicity' (Kennedy 1991, p. 190) evokes the Fall. The very fact that David and Catherine fall from paradise into deviance is a modern reincarnation of the Biblical narrative that illustrates the folly of desire and perversion of humanity.

Furthermore, Hemingway's depictions of androgyny reinforce normative gender positions as both good and necessary for society. Focusing on Catherine as the instigator of deviant gender performance, Hemingway describes her descent into madness as an unavoidable consequence of her departure from the norm, implying that 'far from restoring an archetypal Edenic wholeness, androgyny carries with it the danger of a split personality' (Kennedy 1991, p. 204). Pathologising gender and sexual difference is also evident in the novel's evocation of the 'inversion' (Freudian) and the implied narcissism of the androgynous, and therefore lesbian, female (see Chapter 6 about the space Greek Gold). The early twentieth-century idea of inversion (premised on the idea that non-feminised females wish to be men) is displayed not only through Catherine's physical transformation, but also in her sexual assertiveness and practice:

> Her desire to get a haircut identical to David's marks the onset of her compulsion to become a boy, and once she has short hair, she wants to complete the transformation by assuming the male role in intercourse. In a tantalizing passage Hemingway suggests that, with the help of a sexual appliance, Catherine 'takes' her passive husband: 'He lay there and felt something and then her hand holding him

> and searching lower and he helped with his hands and then lay back in the dark and did not think at all and only felt the weight and the strangeness inside and she said, "Now you can't tell who is who can you?" Even stranger than his inner sensation, however, is Catherine's taking the name "Peter" and then calling her husband "Catherine"'. (p. 203)

Additionally, female narcissism as applied to the pathologising of lesbianism is evident in Catherine's actions. According to Kennedy, 'she forces David to assume her vacated identity as the "girl" and nominally to become her, so that she can make love to herself. Here we witness the first eruption of the narcissism which will later manifest itself in imagery of the mirror at Aurol's hotel' (Kennedy 1991, p. 203). Although the story was written in the middle part of last century, many of the connotative aspects of inversion and narcissism attributed to non-normative female performance continue today. SL's Eden as a space of sexual intent evokes these anxieties in that the performance of the majority of avatars is highly feminised, almost as if averse to any other depiction. Furthermore, in consideration of the potential for male infiltration into the site, Eden can be considered as a space signifying *the* Fall itself. On the margins of the corporeal world, it is a space that can entice male users in to experiment not only with deviant sexual practices but also with deviant (gender-swapping) performance. By embodying a female avatar (and selecting 'female' in the SL set-up phase), males can participate in a feminised performance of themselves. In Hemingway's novel, this altered state is seen as dangerous and is only rectified through David's return to heteronormativity, and from sin, by engaging in a full heterosexual relationship with the Spanish Marita, who is freed of her bisexual tendencies.

Conclusion

The sexually deviant woman is represented both in avatar representation and through the myths and connotations of Eden. Notions of Eden hold that female sexuality is dangerous, particularly in reference to morality and the potential for power. This power, however, can be called into question and reveals itself more as an anxiety of patriarchy than one with any real tangibility. This is because female sexuality is continually reified through the male-gaze, even in the performance and practice of alternative sexualities. Even lesbian sexuality is understood through the male-gaze; through pornographic signification as well as popular culture the male viewer is inscribed as prominent even in his physical and temporal absence. Female identity then becomes subsumed into hetero-patriarchal discourse so that images and representations conform to an acceptable standard. This is particularly limiting for individuals who do not conform, making it necessary for there to be spaces of safety and tolerance. The case of Eden highlights the contradictions within the contemporary time period in which alternative sexualities have become more acceptable; however this acceptance has come at a cost of disavowing certain ways of being. Rather than being a fluid and open space of identity expression, Eden proves to be a space of heteronormative reinscription of the female body.

6
Subverting Gender

Introduction

This chapter explores the female-only lesbian space of Greek Gold. It provides a contrast with the space of Eden, as well as many other sites on SL, due to the diverse representations of female avatars within the space which go against the prevailing gender normativity and hyper-gendered embodiment seen elsewhere on SL. Due to the varied expressions of femaleness shown within Greek Gold, it is an excellent site for the exploration of the ideas of gender fluidity and lesbian identities, particularly that of masculinised female performativity. Drawing heavily on Judith Halberstam's work on the masculine female subject, the analysis of Greek Gold illuminates the fluidity of gender and sexual identity expression, underscoring the premise that gender is not static within the binary of male/female, masculinity/femininity as often accepted within mainstream culture. The virtual environment of Greek Gold is a superb illustration of the performativity of gender as seen in the corporeal, and ultimately subverts the idea that masculine and/or androgynous females want in some ways to be men, thus overturning the notion pathologising gender difference. With the possibilities of embodying a desired or preferred self, the prevailing representation of subversive corporeal gender identities reveals both an acceptance of, and pleasure in, embodying a masculine self as female, thereby removing the default presumption of masculinity to biological maleness.

This chapter discusses the masculine female subject within Western discourse including: the positioning of female masculinity as pathological and threatening; notions of lesbian identity and desire in particular reference to the butch/femme relations; and the subversive displays of gender and sexuality as seen on Greek Gold as queer political practice.

The body as subversive is seen through the representational aspects of identity as well as desire in framing identity as not normative. The discussion of avatar representation on Greek Gold borrows, and expands upon, Halberstam's conception of the masculine female subject as principally a gendered identity before a sexual identity. Separating gender from sexuality removes the popular premise that masculinised females 'perform' a male identity for sexual and relationship purposes, and the assumption of mimicry often associated with butch, androgynous and masculine identities. By separating masculinity from the biologically male, androgynous, butch and non-feminine identities can be viewed as having a desire to perform, embody and identify with gender outside of the rigid binary relating to biological sex. The desired subject position of the corporeal masculine female is not a male self, but an idealised representation of her-self as a reinscribed female.

The sexual female: From ancient Greece to modern society

As noted in the analysis of The Lost Gardens of Apollo, Zeus and Eden, the historical imagination is hailed to make meaning of the virtual spaces. In imagining a nostalgic and mythic past, the spaces are framed by fantasy, desire and escape from the rigidity of the present. Unlike Eden, which arguably draws upon the heteronormative form of woman and female sexuality, Greek Gold selectively remembers elements of ancient Greece, particularly the Amazons, Spartans and Sappho. Whilst much has been made of the ancient Athenian attitude to women as inferior and unable to participate as full citizens, and so being effectively marginalised to the domestic space, these three distinct cultures present important exceptions.

The traditional role of women in both ancient Athenian culture and mainstream Greek myth was generally that of a passive bystander, and occasional helper, to the male hero or dominant figure (Lefkowitz 1986). For imperial women, as well as goddesses, power was seen as an extension of the male, whereby 'the patronal acts of these women were seen as being a part of the extended activities of their male family members, and thus did not violate gender roles or social institutions' (Montserrat 1998, p. 170). It can be argued, then, that invoking ancient Greece and its myths in SL could have the effect of reinscribing this patriarchal function. Whilst this is arguably the case in certain instances, there are also those myths and cultural re-imaginings that resignify female agency and power. One only has to turn to the

TV production *Xena: Warrior Princess* (1995–2001) to find an example of a powerful and independent woman reinscribed into ancient narratives. Although arguably perpetuating some stereotypes in the form of female attractiveness, 'Xena breaks many of the rules of feminine decorum' (Inness 1999, p. 161). Similarities with *Wonder Woman* are also hard to overlook. Although commonly associated with the TV series starring Lynda Carter, which saw a heteronormative portrayal of the lead character, *Wonder Woman* also has connotations with *Wonder Woman Island* (otherwise known as Paradise Island) and the Greek myth of the Amazons: 'Wonder Woman's all-women birthplace provides a haven from patriarchy, as well as a response to it. For Themyscira, the more contemporary name for Paradise Island, is governed by women for women' (Peters 2003, p. 3). Similar to Greek Gold, Paradise Island is a fantasy space that provides a safe place for women within a representative spatiality of a mythologised past.

The historical narratives of the Amazons and Spartans resonate strongly with the SL island of Greek Gold. Outside of Greek society itself, the Amazons are alluded to in Greek myth as a female-only society who fought wars like men and sometimes aided the Greeks in their battles. Ancient Sparta is notable in illustrating different gendered norms and behaviours from those of ancient Athens, with many accounts describing the equality of the society in comparison with most other regions of the ancient world. For the SL space of Greek Gold to draw upon the ideas of Sparta, with particular reference to the roles and behaviour of women, is unsurprising. In addition to the role of women in civic, domestic and military life, there were the values of democracy and equality which were central to the Spartan ethos, as well as strength and courage. Paul Cartledge describes the Battle of Thermopylae – which saw a Spartan-led Greek force see off a large invading army of Persians – as a key milestone in the preservation of the Spartan (and Greek) way of life, thereby enabling 'the development of the civilization that we (as Westerners) have chosen in crucial ways to inherit and learn from' (2004, p. 167). Furthermore, Cartledge (2004, p. 168) describes the importance of the cultural imaginary in view of ancient Western history:

> Another answer to the question why the ancient Spartans matter to us today concerns the impact of what has been variously labelled the Spartan myth, mirage, or tradition. That is to say: the variety of ways in which Sparta and Spartans have been represented in mainly non-Spartan discourses, both written and visual, since the late fifth

> century BC has left a deep mark on the Western tradition, on the understanding of what it is to belong to a Western culture.

It is this 'tradition' that is hailed where the idea of Sparta, its customs and way of life, is revered as a special time and place in our cultural development. It can be viewed as a place of near perfection, replete with freedoms and liberties (to full citizens of Sparta, of course) that have been seemingly eroded by the restrictions and humdrum of contemporary existence.

Social freedoms and liberties rouse a particular response in reference to our contemporary model of normal and deviant conduct. Although many Westerners are afforded a high standard of living that allows for various consumer freedoms, as well as a good standard of health and freedom from risk, disease and harm, there is an argument that such general well-being comes at the price of over-regulation of our private and social lives. In contrast, the sexual freedom of ancient Spartan society, particularly for women, can be imagined as particularly attractive. Some of the better known (as well as imagined) Spartan social practices and customs are those that would today be deemed inappropriate, and even categorised as criminal in some instances:

> institutionalized pederasty between a young adult citizen warrior and a teenage youth within the framework of the compulsory state-managed educational cycle; athletic sports including wrestling practised officially – and allegedly in the nude – by the teenage girls; the public insulting and humiliation of bachelors by married women at an annual religious festival; polyandry (wives with more than one husband each); and wife-sharing without either party's incurring the opprobrium or legal guilt of adultery. (Cartledge 2004, pp. 174–175)

What stands out distinctively in the above examples is an idea of women that is far removed from Athenian discourse, as well as the construction of women from the ancient world through to modernist discourses. Indeed, such behaviours as ascribed to women would even seem outlandish and ultra-progressive today, placing Spartan practices perhaps at a point in the future of our current conservative norms.

In addition to the Spartan tradition of female agency and empowerment is the prominence attributed to the Athenian island of Lesbos, the birthplace of the poet Sappho. The late seventh/early-sixth-century BC poet is renowned as one of the first female writers to include elements of female homosexuality (homophilia) in her work. Although there are questions regarding the autobiographical accuracy of accounts of the poet herself, her poetry is revealing of female conduct and attitudes in the outer regions of the ancient Athenian Empire. Judith Hallett explains that, for later ancient Greek scholars, any homosexual liaisons that may have initially been ascribed to Sappho herself were 'suspended' and mostly overlooked. That she wasn't forgotten altogether was due to her esteemed place as a revered poet in the Greek Athenian tradition. As Hallett explains, such an aversion to female homosexuality can be viewed as hypocritical in view of male homosexual and pederastal conduct (1979, p. 449):

> By comparison, the homosexual liaisons attributed to the male poets of Sappho's time do not meet with similar disbelief or disapproval. A number of these same authorities refer to the homosexual involvements of Greek male lyric poets as established facts; like virtually all ancient testimony on the lives of Greek poets, they do not give the impression that male pederasty, at least for the 'active' partner, was thought cause for shame.

Such aversion to female homosexuality has persisted into recent history, with scholars divided over the moral and aesthetic properties of Sappho's writing, in view of a fascination with its subject matter. Hallett notes a passage by a 1970s scholar suggesting that one of Sappho's passages could be read as a: 'clinical record of acute symptoms suffered by a "masculine lesbian" during an anxiety attack' (1979, p. 450), showing the tendency of patriarchal discourse to pathologise gender and sexuality difference. That Sappho's poetry has survived as having historical and cultural merit despite its supposedly unsavoury subject matter is testament to her position as an esteemed writer from ancient times through to today. Furthermore, she now has an enduring position in the lesbian community whereby her name, as well as the descriptor 'Sapphic', is now attributed to female homosexuality. Despite the debatable accuracy of Sappho's own sexuality, this attachment has persisted (Hallett 1979, p. 451):

> Yet the view that Sappho not merely indulged in, but exhibited an exclusive preference for, homosexual acts has only gained widespread currency in the past few years. Indeed, the modern sense of the words 'Sapphic' and 'Lesbian' is largely responsible for popularizing the view of Sappho as an exclusive, and physically practicing, homosexual.

Of course, we cannot avoid reading such texts from our own position in history, and therefore Sappho's poetry is difficult, if not impossible, to scrutinise outside the lens of modern patriarchy and contemporary culture, specifically in relation to its homosexual connotations. From a lesbian-feminist viewpoint, Sappho appears to be a lone voice speaking out against the prejudices of an ancient and persistent patriarchy, where female eros flourished but was forbidden to be spoken in the historic record.

It is unsurprising that lesbian spaces re-imagine Sappho and Lesbos as idealised places of female sexuality. On Greek Gold, the Mediterranean island environment is evoked in its spatial and representational characteristics: idyllic beaches, gardens and waterfalls exist in a climate that is eternally summer. Similarities with island tourism are extensive in the idyllic island landscape that symbolises freedoms not normally associated with the everyday. Moreover, there is a freedom of expression and desire on the island, reminiscent of that suggested in Sappho's poetry, where women are able not only to possess sexual feelings, but also to act upon them. Hallett explains that the desire expressed by Sappho contradicts the position of women as passive, being 'conditioned to view as unfeminine any woman who openly expresses sensual attraction for another human being; she is taking the "sexual initiative" and behaving as only men are supposed to' (1979, p. 455). Furthermore, 'when that object of allure is a woman, as in Sappho's poems, the aggressive female is considered doubly masculine' (Hallett 1979, p. 455). Greek Gold both challenges assumptions regarding 'proper' female sexual behaviour and embraces the masculine qualities of sexual assertiveness. In doing so, it adds to the reworking of masculinity as performed within the space, as a welcome and positive attribute for females both to possess and express.

The masculine female subject

In the modern era, gender was strictly controlled to produce a strict binary between the male/masculine and female/feminine. This was policed through religious dogma and further reinforced through the concept of a naturalised biology as determined at birth. Modernist discourse disregards the Spartan and Sapphic views of gender and sexuality as both heretic and irrational in consideration of the broader modernist aim of rational categorisation and hierarchy. Such rigid gender characterisations have been upheld through continual cultural reinforcement that ascribes deviancy to gender fluidity. Although sparse, there is some scholarship regarding the non-feminised subject that emerges in different guises in recent history. Judith Halberstam, in her work regarding masculinity in women, argues that the non-feminised female subject is not a product of recent politics of feminism and queer rights movements, but instead has existed 'for at least two centuries' (1998, p. 45). Importantly, she distinguishes various forms of female masculinity from the overarching term 'lesbian' to describe non-feminised subjects, revealing the complexities of both gender and sexuality to identity, and emphasising the role that female masculinity has had on the development of masculinity throughout history (Halberstam 1998, p. 48):

> [T]he momentous negotiations about gender that took place at and around the turn of the [twentieth] century, which were created by earlier developments, produced particular forms of femininity and masculinity and clearly showed that femininity was not wed to femaleness and masculinity was certainly not bound to maleness. The transition from affiliation marriages to romantic marriages, the development of the women's rights movement, the trials of Oscar Wilde, the social upheaval caused by World War I, and the development of sexological models of sexual definition all played a part in untangling once and for all the knots that appeared to bind gender to sex and sexuality in some mysterious and organic way.

This analysis is instructive in providing a historical foundation for the multiplicities of gendered forms of contemporary identity, illustrating that recent forms of gender are not ahistorical and that gender fluidity is something that has always been present in society, albeit in different guises than those we recognise today.

Halberstam (1998) makes an important distinction in separating gender from sexuality which is a crucial step in the historical account of the construction of masculinity and femininity, and is illustrated in the conception of the contemporary androgynous and masculine subject. She views historical gender as a distinct set of power relations, justified by biological sex characteristics cemented in the binary of masculinity/femininity, through which all individuals are characterised, even in their deviancy (Halberstam 1998). It is through this strict categorisation that deviant gender (and sexual) deviances were made to operate, producing such subjects as the 'tribade' and the 'female husband', as well as female romantic friendships that may suggest more than the asexual relationships some historians have assigned to them. The absence of a historical sexual record of female gender and sexual deviance is part of the greater denial in society of female sexuality as a whole. Marcius Vicinus (1992, p. 470), a lesbian historian, suggests that 'lesbian history will remain a history of discontinuities', explaining that:

> we rarely know precisely what women in the past did with each other in bed or out, and we are not able to reconstruct fully how and under what circumstances lesbian communities evolved. Our history includes teenage crushes, romantic friendships, Boston marriages, theatrical cross-dressing, passing women, bulldykes and prostitutes, butches and femmes, and numerous other identifications which may – and may not – include genital sex. When we can't even claim a specific sexual expression as a key to our past, we must accept a fragmentary and confusing history.

Halberstam, however, sees this as less problematic, arguing that there can be an implicit assumption of sexual deviance given the historical presence of the masculine female subject. Furthermore, she contends that the category 'lesbian' may be itself problematic because 'sexual identities, when and where they emerge as identities, tend to be exceedingly specific and often refer to a limited range of pleasures' (Halberstam 1998, p. 56). Therefore, androgyny and the masculinised female subject may be explained more accurately in the terms ascribed to a historicity of gender than to the sexual category of lesbian that 'has conventionally come to be associated with the asexual, the hidden, the "apparitional" and the invisible' (Halberstam 1998, p. 56).

Halberstam traces the masculine female in Western society from the pre-twentieth-century and pre-medicalised sexual discourse through to contemporary representations. Although she describes the masculine females as having same-sex desires, she importantly differentiates same-sex practices from identity politics, arguing that pre-twentieth-century manifestations of gender were subsumed within the masculine-feminine binary which lay outside of the markers of sexual identity that we have today. This has the effect of placing gendered practices outside of our contemporary culture and importantly allows for a re-construction of gender and sexuality, illuminating the development of current representations of non-feminised females in society and culture as being part of an on-going development rather than a recent phenomenon dependent on the social conditions of the last 30 years. In the pre-twentieth-century literature there is the appearance of the tribade and the female husband that are both incorporated into the category of normative binary gender that assumed an incomprehension of female same-sex desire. By the turn of the century, and in a large part due the medicalisation of sex and sexuality, such practices became acknowledged, allowing for the development of sexual identities (Halberstam 1998, p. 76):

> When the idea of sexual identities did come to dominate people's thinking about sex and gender, it was not some idea of an autonomous lesbian desire between women or a notion of outward hermaphroditism that provided the basis of those notions of identity; it was gender inversion.

Halberstam argues that the motivation of sexological study and categorisation was 'the desire to reduce sexuality to binary systems of gender difference' (1998, p. 76) that understood sexuality as an adherence to, or refutation of, normative modes of masculine and feminine behaviours. Female inversion, then, was assumed to be a desire for some women to achieve masculinity in a society ruled by men (Halberstam 1998, p. 78), rather than an active desire by women for women and the display of behaviours which may have been masculine, but which could also be defined outside of the realm of maleness. In addition, the inversion model did not allow for a range of gender or sexuality identification, in an effort 'to force multiple expressions of sexual and gender difference that they (the medical profession) were absolutely committed to bolstering and preserving' (Halberstam 1998, p. 83). This binary model of identification largely remains in place today, as can be noted in the strict policing of gender performance enforced through societal apparatuses (such as public toilets and restrooms) and through the gaze, and active judgement, of the majority.

The binary model still permeates contemporary notions of acceptable queer identity where often it is assumed, by the majority as well as in some sections of the queer community, that a highly masculine female wants to be a man and may be in some sort of pre-transsexual gender crisis. In addition, it is often assumed that any female with masculine behaviours is either a lesbian, wanting to be a man, or both. Halberstam (1998, p. 96) illuminates the complexities of identity for masculinised females in the last 50 years, where lesbian sexual identity became privileged over an ambiguous gender identity to the point where gender fluidity and notions of inversion became demonised:

> As the notion of lesbianism gathered strength, so the masculine woman became a paradoxical figure within lesbian communities; she was representative of those communities as the 'butch', but she was

> also ultimately rejected as an anachronistic reminder of the rejected discourse of inversion. Indeed, to this day, many contemporary lesbian communities signal their modernity by denying the stereotype of the mannish lesbian.

It is possible that this disdain for female masculinity has roots in the gender wars as prompted by the initial movements of feminism in the 1970s where men, and associated masculine behaviour, were heavily criticised (particularly in the radical feminism of Andrea Dworkin and Catharine MacKinnon) and seen as detrimental to the female gender as a whole. In more recent history, and as a response to the feminist movement's rejection of men and male behaviour, the non-feminised female was made an object of ridicule by both men and those who did not wish to be aligned with the more radicalised and ultra-conservative positions of the feminist project. Indeed, in an arguably post-feminist world that supposedly privileges subjective identities that the masculine female represents, the subject remains problematic, marginalised and delegitimised as an actual and proper identity.

Whereas representations of feminised males have become more acceptable in Western society, as can be seen through various television shows of the last ten years, such as *Glee* (2009–) and *Modern Family* (2009–), the masculine female remains marginalised both through under-representation and/or ridicule in the same media forms. Sue Jackson (2009, p. 201) describes these media representations as an extension of the commodification of the female for the male gaze:

> As commodity, she mirrors her heterosexual sister, designed within cultural norms of heterosexual attractiveness and stamped with sex appeal, commonly referred to as the 'luscious', 'hot' or 'lipstick' lesbian.

For instance, the lesbian drama *The L Word* (2004–2009) features an almost entirely feminine cast. The two masculinised characters of the show (Shane and Moira) both problematise depictions of their masculinity. Moira is represented as so masculine that she ventures on the obvious path of trans-gendering (becoming Max Sweeney). Conversely Shane, who is androgynous with a highly stylish appearance, which tempers any mannishness, appears as a playful and non-threatening boyishness. Claudia Card remarks of the show that 'despite its obviously good-willed attempts at representing diversity, the show constructs lesbians in ways that mimic other media constructions of the ideal American woman', and that 'nobody looks like a dyke (despite Jenny's haircut), not even Shane (anymore)' (2006, p. 227). Within heterosexual depictions of masculinised females, often described as 'tough girls', a similar process is undertaken that essentialises the female gender. In looking at 'tough girls' on screen, Sherrie Inness (1999) describes the way that women are presented as tough in such films as *The Silence of the Lambs* (1991), *The X-Files* (1998) and *Alien* (1979), but that this toughness is ultimately devalued by a repeated emphasis on the main characters' femininity (Starling, Scully and Ripley) in relation to the male characters on screen. She states that 'the same shows that depict positive messages about the ability of women to be tough and capable also suggest that women cannot be as tough as men' (Inness 1999, p. 101), reinforcing the gender binary and ensuring that the privileged masculine position remains the property of men.

It is the very threat that the masculine female possesses in reference to male masculinity, particularly regarding the privileges of male sexuality and the phallus, which renders the identity as so problematic. This is applicable both within the lesbian community and outside of it. Within the boundaries of lesbian feminism, masculinity is too readily attributed to a female subjugation based on violence and control, in everyday social life as well as sexual encounters. Therefore, an anxiety occurs around the expression of masculinity whereby the masculine is indeterminable from an often oppressive maleness as experienced historically and culturally. Although, as Halberstam notes, 'in contemporary lesbian culture, there are still many women whose desire works through masculinity and through phallic fantasy and through sexual practices that phantasmically transform their female bodies into penetrating male bodies' (1998, p. 72), there are often anxieties within lesbian communities that see symbolically masculine sex practices, as well as appearances, as highly problematic due to the associations with male control and violence. The heterosexual culture conversely equates the threat with undermining male power, particularly at the site of sex, rendering maleness itself obsolete. However, by reinforcing the necessity of the phallus (with the assumption that masculinised lesbians are males without penises), the patriarchal primacy of phallic power can be maintained, rendering the position of the masculinised female as undesirable and prone to ridicule as an inappropriate, and threatening, other.

The island of Greek Gold

In opposition to mainstream culture, there is both an explicit acceptance and implicit approval, of masculine and androgynous avatars on

Greek Gold (masculine female avatars are often referred to as butches or bois on the island). Alongside the welcome notes and advertising that emphasise acceptance of butches and bois, such representation is easy to find with many avatars appearing androgynous (having both feminine and masculine qualities). There is a fluidity of gender throughout the space from the very feminine to the very masculine, with some avatars even having the appearance of trans-genderedness (although, as noted by Halberstam, one shouldn't assume trans-gender aspiration for females who appear highly masculinised). As shown in the image below (Figure 6.1), highly masculinised representation is welcome on Greek Gold, including an almost entirely male appearance. Note that the avatar on the left is wearing 'pasties' (plasters over the nipples) that conform to the 'no nudity' regulation of the space. This example is representative of the complexities of the masculine female residing outside of female norms, including those relating to the modest portrayal of the body. Avatars such as this illuminate the paradox of masculinity in a queer, especially female-only, environment. Although maleness itself is disallowed in the form of CL males, the appropriation of masculine qualities is seen as virtuous, subverting the patriarch by queering it. Also, many instances of highly masculinised avatars do not have the appearance of heterosexual males; they seem to lie on a continuum of androgyny through a queer idiosyncratic appearance. Whereas such representation may cause confusion in a heterosexual environment, there is little ambiguity in a queer space whereby such markers are easily understood and interpreted. It is perhaps for this reason that the Greek Gold space appears less prone to the infiltration by males into a female-only environment (although it is impossible to tell for sure). Due to the practices of signification and marking of the body in queer communities, it is likely that avatars not displaying queer markers may be identified as imposters.

It is common to see highly feminised avatars alongside androgynous and masculine females, and within the research timeframe there was always a near equal representation of gendered avatars on the continuum of highly feminine to highly masculine. Unlike the site of Eden, not present when visiting Greek Gold were hyper-gendered and hypersexualised avatars in large numbers. Hyper-gendered avatars were often seen in close proximity with androgynous, butches and bois, therefore queering the depiction of hyper-genderedness as understood in mainstream culture. In addition, many feminised avatars often had markers of queerness displayed to further contextualise their identity within the space. Alison Eves (2004, p. 494) in her research into butch-femme

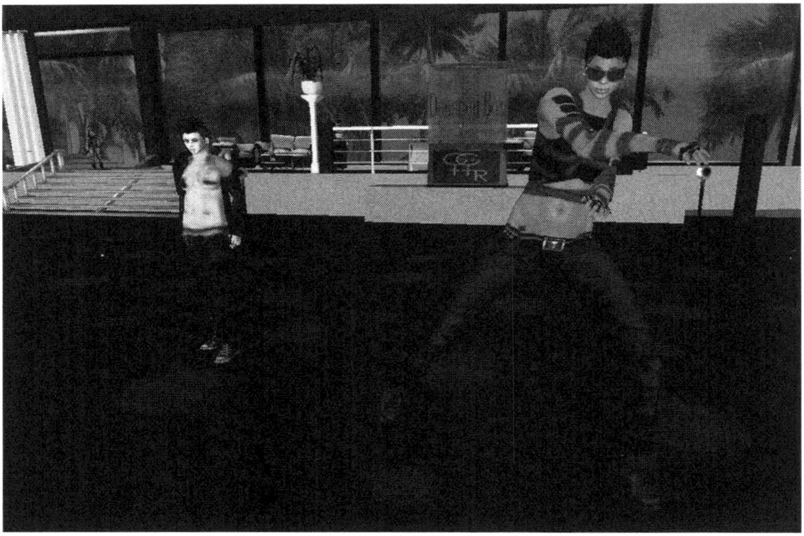

Figure 6.1 Dance-floor, Greek Gold
Source: With permission from Linden Lab and the residents of Second Life.

identities, describes the subject position of the femme in queer spaces and the phenomenon of 'femming-up' in queer space:

> Butch/femme can be seen as shared subcultural resources which are used to construct alternative lesbian subject positions. There can be no 'outside' of gender so some femmes chose 'femininity with a twist'. However many lesbians, and even butch and femme women who are more familiar with the subcultural styles, have difficulty distinguishing between femme and heterosexual performances of femininity, particularly away from lesbian spaces. This is in spite of the hyper-femininity in much femme style and femmes' insistence on the knowing and parodic intention behind this .... The drag-like quality of femme gender performance was confined to gay space for some femmes. On a day to day basis ... femininity was played down and was described as being much closer to and probably indistinguishable from the femininity of heterosexual women. Gender performances are read in particular contexts, which have a bearing on their meanings regardless of intention.

The phenomena of 'femming-up' can be read as a contextual play on sexual power as experienced within queer space, with the threat of male sexual violence removed. The femme is able to experiment with various aspects of femininity that may attract unwanted attention in heterosexual space, with the connotations that she somehow wants male attention in that environment. Through the performance of femininity in queer spaces, hyper-gendered and hyper-sexualised femaleness becomes a subject position of power; she is free to perform gender and sexuality without fear of sexual violence that is prevalent in heteropatriarchal environments.

Economy and power

It is necessary to make explicit the role of economics to representation and gender identity on SL in general, and specifically within spaces such as Greek Gold. SL is highly reflective of corporeal Western society due to its economic system, and although there is a far greater acceptance of difference on the platform, much of it has referents to the mainstream capitalist society from which it is derived. For this reason, there are important consequences for membership within the wider SL community such as the ability to perform a self, or selves, in the environment. Much of an individual's ability to represent an idealised self stems from their ability at mastery in-world (the amount of time they are able to allow themselves to interact and explore SL spaces), as well as their financial capability (the amount of Linden dollars they either transfer from CL money, or conversely make in-world through market transactions). Agency is directly related to an individual's ability to embody, present and act within the SL environment. For instance, an individual new to SL who has no Linden dollars would have to use a default avatar as provided in the signing-up process. Given the limited options of identity, and the adherence of the given avatars to CL heterormativity, there is little scope for gender subversion – as of March 2012, there were seven 'boi' options and one androgynous option for complete avatars in the SL marketplace as opposed to 1,582 male and 3,700 female (Second Life marketplace 2012b). The economic resources required for gender deviance resonates with the late-nineteenth-century conditions for the aristocratic subversion of gender (Halberstam 1998), as well as some more contemporary allowances afforded to those who subvert gender norms but are permitted this deviance due to economic power (Figure 6.2).

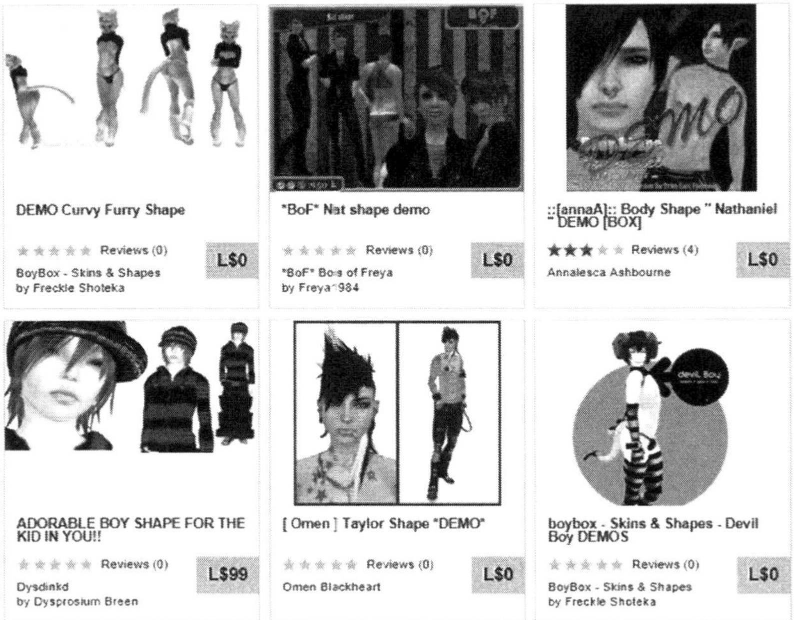

Figure 6.2 SL Marketplace, search query: 'boi'
Source: With permission from Linden Lab and the residents of Second Life.

Historically, masculine females had some ability to perform identity outside that of 'passing' as men if they were from the aristocracy or gentrified classes. The ability to live outside of the conventional marriage arrangement was only tolerated when, particularly as women, they had the ability to support themselves: 'Masculine identification with social impunity required money and social status' (Halberstam 1998, p. 87). In contemporary Western culture, this economic imperative is repeated, in that those who have money also have access to identity outside of the norm. In a study analysing attitudes to homosexuality in 35 countries, Robert Andersen and Tina Fetner found that tolerance, whilst somewhat dependent on the overall affluence of a nation, is more directly related to an individual's social class within that nation (2008, p. 955):

> It was found that tolerance for homosexuality is much more likely among professionals and managers than among the working class ... The results further demonstrated that class and national prosperity

> interact in their effects on attitudes. In other words, how national economic factors influence an individual's attitudes depends, at least in part, on the individual's position in the economy.

So, tolerance is afforded to homosexual individuals who are part of the more privileged classes in society due to their participation within that social group. Similarly, for those who are not in affluent social positions, representation as an 'other' or suspected deviant is limited: 'Although the middle classes become more tolerant of homosexuality as economic development rises, the working classes have similarly less tolerant attitudes regardless of the level of economic development' (Andersen & Fetner 2008, p. 956). This can have the effect of normalising representation for those who have limited economic resources in order to reduce harassment, violence and discrimination.

For lesbians and females with a non-normative identity, this effect can be more pronounced due to their position as an already economically marginalised group. The reduced economic capacity of women as a whole can further impact on the ability of females to represent an identity outside of the conventional. This can manifest itself both corporeally and within the virtual environment. For instance, it is noted by Dotson, Hyatt and Thompson that lesbians generally make less money than gay men, reducing their agency in Western capitalist societies, and ultimately reducing representation in the mass culture. In their study looking at marketing to homosexuals, they outline the economic disparity between lesbians and gay males: 'Their median household income is also lower than that of the gay male population. This is true even though 77 per cent have college degrees and 54 per cent have managerial occupations' (Dotson et al., 2009, p. 433). Accordingly, female access to technology can also be limited on a global scale, with far fewer women than men having access to the internet and therefore limiting their access to participate on platforms such as SL. Nancy Hafkin and Sophia Huyer (2007, p. 26) assert that:

> Despite the lack of gender-specific quantitative data, project-level qualitative data have established that ICTs are not gender neutral. ICTs impact men and women differentially, and in almost all cases, women have many disadvantages that result in their having less access to the technology and therefore less use of it.

Additionally, within the economy of SL itself, lesbians are marginalised economically, with non-feminised avatars, clothing and accessories being difficult to find, therefore making non-normative representation difficult. There is a reliance on perseverance and community to overcome this problem, which is related to the overarching factors of in-world mastery and financial ability. Simply searching for 'androgynous avatars' in the SL Marketplace for instance, produces few results – it is easier to embody a dinosaur or other creature than an androgynous avatar. It is only with a degree of familiarity with the in-world environment and LGBTI community that an individual can find stores that facilitate non-normative gender representation. *Oh Boi Magazine*, an e-publication for lesbians, highlights the problem of finding non-feminised clothing in SL. This problem in market identification could be due to the absence of CL marketing to lesbians and females who wish to display a non-feminised appearance. The fact that there is little to model on in terms of CL clothing limits the creativity of the marketplace vendors, which is probably in turn reinforced by a reduced market in terms of lesbian and further, masculine female, identification.

Lesbianism, the masculine female and butch/femme desire

Many corporeal Westernised inner cities allow for fluid gender expression as well as diverse sexualities, such as the gay ghettos described by Valentine and Skelton (Second Life Marketplace 2012b), and the 'queer-friendly neighbourhoods' defined by Gorman-Murray and Waitt (1998). Greek Gold presents multifaceted lesbian representations, and so can be seen as an extension of this urbanised queer identity, particularly in relation to the expression of fashion and bodily adornment. Representations include highly feminised avatars (like those seen on Eden), as well as very masculine avatars that are defined in the space as female even in the absence of 'proof' of some essential femaleness often signified by female genitalia. The very absence of such proof on Greek Gold, as is assumed within the physical environment, acts to destabilise the conditions of gender associated with sex and sexuality, as well as the very foundations of binary gender and sex performance. The fluidity of masculine and feminine performance on Greek Gold illustrates a detachment of an assumed naturalised gender, from biological sex characteristics and the very conditions of biology and nature. Consequentily, sexual preference is also removed from any type of natural identification and causation; attraction and intimate interaction on Greek Gold occurs under the presumption of femaleness; however, the gender category of female is destabilised through various performances of androgyny and masculinity undermining the notion of femininity as 'naturally' belonging to the female. Both gender and sexuality, then, are reconstructed as separate, albeit related, modalities of performance under the prevailing assumption of fluidity that dissolves the binaries associated with biology and nature.

If the biological sex of individuals behind the avatars is assumed to be female, then the variety of gender performance and sexuality as seen on Greek Gold can be assumed to belong to various categories of femaleness. Interaction and attraction are presumed entirely female regardless of the appearance or visual modality of the embodied avatar represented on the screen. The diversity that is present is evidence of the need to view gender as a fluid state, far more complex in scope than is accepted in mainstream understanding. The very ideas that are perpetuated around masculinised females wanting to be men are entirely subverted through the functionality and capabilities of screen embodiment; there would be no point in embodying an androgynous or highly masculinised female in an environment where one is free to embody a physique of their own desire, such as on SL. If one wished to be male or to delineate oneself by the binary structure of essentialised gender, the various regions of SL would provide enough scope to do so. Instead, within the space of Greek Gold, there are many individuals who embody female avatars that are either androgynous or masculine, which actively subverts the gender binary of mainstream culture. Females within this space do not wish to be men; rather it can be assumed that they acknowledge their femaleness (in terms of a gendered identification) in addition to destabilising the very conditions that this gender identification is founded upon, namely the historical, cultural and political categorisations of behaviour and performance. What further undermines the gender binary, as well as many of the assumptions about female inversion and the presumed transgendered identity of masculinised females, is the sexual dynamics of butch and femme subjects. Butch/femme desire reveals the complexities of identification within the category female and, rather than reinforcing a performance of heterosexual desire as argued by many feminists in the 1970s, it in fact subverts the foundations of normative desire and sexual practice.

Judith Roof contends that by rejecting 'the oppressive patterns of heterosexual relations', where heterosexual relations are assumed to be the basis of butch/femme desire, 'it overvalued a gender essentialism that in the end only reified the very system it wished to critique' (1998, p. 27). In addition to the various movements that have led to a reformulation of sexual and gender identity categorisation – such as trans, intersex and SM identities – the butch/femme identification has, since its rejection by 1970s feminists, achieved recognition as a destabiliser of normative gender and sexual roles. However, and as acknowledged by Roof (1998), Halberstam (1998) and Alison Eves (2004), the butch-femme is still considered transgressive in mainstream society, as both threatening to the established patriarchy and as politically regressive by some feminists and conservatives. For instance, Eves (2004, p. 493), in discussing lesbian presentation and aesthetics, summarises the issue of appearing non-heteronormative:

> Negative stereotypes about lesbians in contemporary popular culture have included assumptions about the way that lesbians look and why. The stereotypical image has been butch, and the assumptions have been shaped by discourses of inversion, so that an active sexual desire for women has been seen as necessarily masculine.

There still remains a prevailing attitude that androgynous through to masculine females, particularly lesbians who do not identify as femme, are deviant to the point of inviting violence or threatening behaviour, and an exclusion from participation in mainstream society.

Inversion, in both gender and sexuality as defined by the early-twentieth-century sexologists, is not a satisfactory explanation for the embodiment of a female avatar with masculinised appearance, behaviours and sexual desire to participate in a relationship that can be defined as butch-femme. If butches wished to reject their base femaleness in favour of maleness then the virtual environment provides a perfect platform for a complete transgendered embodiment; for this to be true there would ultimately be no representations of butches, bois or masculinised females within SL's spaces such as Greek Gold. As evidenced through the virtual embodiment of a masculine, but female, form, masculinity is removed from its referent to maleness and a reduction to biological sex as a determination of gender performativity. In her analysis of the corporeal stone butch identity, Halberstam critiques the suggestion that butch-femme desire is a role-play of heterosexual relationships, where 'role players were seen as insecure, immature or unevolved' (1998, p. 131). Instead, the butch and femme represent an attachment to their preferred gender performativity as experienced in behaviour, appearance and a preferred desire towards an opposing gender performativity. It is required, as argued through Halberstam's work, that gender is seen as separate from sexuality, which in turn is necessarily separate from biological sex. In fact, in determining an appropriate language for discussing sex, sexuality and gender, the vocabulary seems lacking (Eves 2004, p. 483). This appears to be an issue that Eves (2004, p. 487) encounters herself when trying to explain the basis of desire within butch-femme partnerships, reducing it to the discourse of heteronormative binary:

> Femmes linked sex and gender in a conventional way but challenged the connection between gender and sexual orientation by being attracted to women, although where this is or is understood to be for masculinity it can also be recuperated by the heterosexual imaginary. Butch accounts disrupted the link between sex and gender, but as masculine their attraction to women can be seen as conventional, particularly when it is for feminine women.

As shown in the above passage, there is a problem in attaching identity within queer communities to aspects of biological sex and the associated assumptions of sex categorisation (or innate sex characteristics), and the relationship of this 'base' gender to performative gender, sexuality and desire. Halberstam elaborates on this point in her assertion that 'Female masculinity within queer sexual discourse allows for the disruption of even flows between gender and anatomy, sexuality and identity, sexual practice and performativity', revealing 'a variety of queer genders, such as stone butchness, that challenge once and for all the stability and accuracy of binary sex-gender systems' (1998, p. 139). This inadequacy of the dominant cultural language is further illustrated in conceptions of desire within butch-femme relationships, that have often been essentialised in view of the prevailing discourse of gender and sex binaries and the presumption of inversion.

Joan Nestle, a self-identifying femme, explains that the butch-femme partnership draws on the political subject position of the lesbian in the 1950s and rejects the notion of heterosexual coupling mimicry (1992, p. 542):

> Butch-fem relationships, as I experienced them, were complex erotic and social statements, not phony heterosexual replicas. They were filled with a deeply lesbian language of stance, dress, gesture, love, courage, and autonomy. In the 1950s particularly, butch-fem couples were the front-line warriors against sexual bigotry. Because they were so visibly obvious, they suffered the brunt of street violence. The irony of social change has made a radical, sexual, political statement of the 1950s appear today as a reactionary, non-feminist experience.)

Writing in the 1980s, Nestle often felt defensive because of her femme appearance, which was read by many feminists at the time as a victimised embodiment of the conditions of patriarchy rather than a reification of those conditions. Clare Hemmings suggests that butch-femme desire can be read by the way that it offers a repudiation of heterosexual culture: that it is a conscious choice of cultural repudiation in conjunction with an unconscious sexual-object choice, and that this desire necessarily operates on the conscious level for both lesbian, as well as bisexual, femmes: 'What I want to suggest is that a bisexual femme's unconscious repudiation constitutes her as a sexual subject, and that her conscious repudiation constitutes her as a cultural subject' (1998, p. 100). So, within the butch-femme identification, there is a desire and sexual attraction that can be categorised along the feminine-masculine continuum which is below the level of conscious choice, and although that desire may be gendered it should not be reduced simply to a biological sex categorisation. In conjunction with unconscious desire, there is a conscious level of attraction that works at the site of culture. The cultural site of desire is where the butch-femme comes into its own, and is where performativity and attraction work at a level of complexity so as to queer the established conditions of heterosexuality. In describing her own experience as a femme, Nestle (1992, p. 545) argues for an acknowledgement of context within the interplay of butch-femme:

> Make-up, high heels, skirts, revealing clothes, even certain ways of holding the body are read as capitulation to patriarchal control of women's bodies. An accurate critique, if a woman feels uncomfortable or forced to present herself this way, but this is not what I am doing when I feel sexually powerful and want to share it with other women. Fems are women who have made choices, but we need to be able to read between the cultural lines to appreciate their strength. Lesbians should be mistresses of discrepancies, knowing that resistance lies in the change of context.

It can be said that the very subversiveness of the desire between both the subject and object positions is what makes the butch-femme such an enduring site of pleasure for both the butch and the femme. Pleasure is at the site of subversion, or the queering of subject/object interaction and sexual practice, so that gender and sexuality are reified contextually within the interplay of both positions.

It is important when studying gender representation in lesbian environments to note that gender identity within social relationships, intimate relationships and internally in the individual are not static (although it can be assumed that varying degrees of gender play occur in other sexualities). In addition to a gender fluidity that can be seen on Greek Gold as well as other queer spaces, there is a gender identity that is relational and dependent on the environment, including the space itself and its explicit or implicit queerness, and the various individuals who inhabit the space. As noted by Valentine in her discussion of lesbian time and space, lesbians often have to assume a normative gendered and sexual positioning in relation to the heteronormative world they must participate in (Valentine 1993). However, there is also a relational genderedness that is performed between females within lesbian spaces, and between prospective and actual partners, that is continually negotiated. Eves (2004), in her discussion of butch and femme lesbian identities, describes how, when in queer spaces, some butches and femmes play with their gender, such as femmes 'femming-up' to perform a hyper-gendered subject. The performance of gendered identity is dependent on the space that an individual is in, so femmes would almost certainly feel more comfortable performing a hyper-gendered and hyper-sexualised identity in a queer space that minimises the potential for violence. Similarly, within Greek Gold, it is possible for individuals to 'test' their masculinity or femininity, and even play with certain aspects of their sexuality in relation to alternative forms of genderedness to those of their corporeal selves. For example, they may feel more authentic as an individual embodying a masculine avatar on Greek Gold in opposition to a corporeal self where they may feel obliged to adopt a feminine appearance.

Greek Gold as a political (queer) space

The queer body in its representation of difference and otherness is a necessary political project. Although it may be argued that in a postmodern, highly commodified and globalised world, society has become used to subversion as a form of entertainment through the screen, it remains a fact that the very lived conditions of individuals continue to be subsumed within normative categorisations of who belongs and who doesn't, within the binary of normal and deviant. Gender deviance is policed within corporeal society through continual reinforcement of gender separation and supposed gender difference as seen in many pseudo-scientific studies, books and papers on biological sex difference (Fine 2010). Additionally, individuals who digress from gender norms can experience harassment, violence and discrimination in many aspects of their lives. These cultural regulations of gender normativity have a similar effect in virtual communication environments. As evidenced in their study of androgynous avatars, Kristine Nowak and Christian Rauh found that androgynous avatars are received with less certainty and trust online than those whose appearance is clearly masculine or feminine. (2008, p. 18) The authors make a link between anthropomorphism (being more human-like) with a level of genderedness in terms of credibility (Nowak & Rauh 2008, p. 23):

> These results reinforce the conclusion that perception of the avatar has influence on perceptions of the partner. Users should choose their avatars as carefully as they monitor their behavior. Simply put, these results suggest that wrong avatar can make you, literally, look bad, while using a more credible, more anthropomorphic, less androgynous avatar (whether very masculine or very feminine), will make you appear more credible.

This link between androgyny and the 'less human' reinforces the conception that unequivocally gendered forms, even in virtual environments, are seen as more worthy within society and therefore will be rewarded as a result. The alternate side of this equation sees androgyny as a sub-human attribute that should not be afforded the same levels of worth, agency and acceptance as those who conform to binary gender representation.

This supposed threat of androgynous and masculine femaleness can be read as the anxiety about a masculine essence as a male-only attribute. If '(m)asculinity in this society inevitably conjures up notions of power and legitimacy and privilege' (Halberstam 1998, p. 2), then the destabilisation of the discrete attachment of masculinity to male biology undermines the power, legitimacy and privilege afforded to men by virtue of their biological sex. Instances where masculinity is not seen as a discrete and necessary attachment to maleness, as in the case of androgyny or female masculinity, are ultimately threatening to the established order of society. More so than feminised depictions of lesbianism, which can be seen as only mildly threatening to gender assumptions and even included within the boundaries of male fantasy, the masculine lesbian body is highly deviant and dangerous. Barbara Creed (1995, p. 101) suggests that:

> Although ... deviant tendencies are present in the female body, it is the ideological function of the lesbian body to warn the 'normal'

> woman about the dangers of undoing or rejecting her own bodily socialisation. This is why the culture points with most hypocritical concern at the mannish lesbian, the butch lesbian, while deliberately ignoring the femme lesbian, the woman whose body in no way presents itself to the straight world as different or deviant.

So often pathologised as a pseudo-male, the masculine lesbian subject is continually subsumed within the boundaries of the binary sex system of culture in that the masculine subject is interpreted as a male mimicry. It is this assumption that makes a site such as Greek Gold so powerful in the refutation of the system. Representations of maleness on SL, as can also be argued to be the case with many virtual representations of gender, are all pseudo-male. When one embodies a male avatar, that avatar has no biological sex. It is gender performed as an extension of the culture and performative assumptions of the individual behind the representation. Accordingly, representations of masculinity on Greek Gold are performed gender as, it can be argued, is gender within the corporeal environment. For there to be a binding of biological sex to gendered representation, if there were a desire to 'fit' into the 'normal' classification of what is right in terms of normative sex and gender, virtual gender representations would reflect the binary assumed to be natural and self-evident within the individual.

It is also not sufficient to label a site such as Greek Gold – a space for sexual transgression and deviance set aside from mainstream culture – as some sort of playground for the perverse where desire is assumed as abject in reference to 'normal' desire and relational heteronormative fulfilment. That desire is often understood through heterosexual partnerships is described by David Buchbinder (1998, p. 22):

> Female desire in a patriarchal culture is constructed as dependent on male desire, much as a satellite is dependent for its orbit upon the gravitational force of a larger planetary body. The possibility that female desire may be independent of male desire is threatening for men to contemplate in such a culture, for if female desire resists the dominance of male desire, it also resists both implicitly and explicitly, the patriarchal structure and the way in which this locates woman as subordinate to man.

Greek Gold is not a pornographic environment. It refutes the positioning of it as a sexual space based on abject desire, and so is quite dissimilar to Eden. As a lesbian space of diversity that is not based on sex,

Greek Gold rejects the assumptions that lesbians (and other females) use the site for sexual purposes only, and so refutes the idea of lesbian sex as animalistic (Creed 1995, pp. 96–99), as well as the notion that the space is an extension of hetero-patriarchal fantasy (which could be assumed due to the diversity present). Instead, it offers a space of community and identity actualisation that is both safe and non-threatening. Practices of looking are reconfigured in the absence of the male-gaze. Power relations are overturned in the interplay of feminine and masculine–female subject positions. Without the threat of harassment or violence, sexual gendered positions are reified so that sexual, cultural and political power can be both feminised and masculinised, and the positions often passed between.

I consider Greek Gold a political space, not because it demarks itself as such, but because it both recognises and promotes queerness which undermines the conventions of heteropatriarchy. As a leisure space isolated from normative environments and closed to biological males (although there is always the potential for this to be circumvented in a platform that does not verify gender), it could be assumed that Greek Gold would have little impact as a political project. However, it is perhaps because of its unassuming acceptance of all female avatars and its tolerance for gender subversion in an environment that is itself unthreatening and not sexualised within the context of SL platforms, that it achieves a goal of great importance to the queer community. There is a need for the lesbian and queer female body to be a political project. It should not be homogenised to fit within the established conventions of what is normal. Such a project is essential not only for this community, but also for females as a universal gendered category that is still devalued and marginalised within all societies. Greek Gold is an extension of the desire to fulfil multifaceted gender identities that are not limited to biological sex; the referent of sex to gender is abolished through the representations of androgyny and masculinity performed by femaleness. Masculinity to femaleness is not the same as masculinity to maleness. The notion of inversion is reductive and regressive and devalues the complexities of gendered experience.

Conclusion

The differences between our virtual screen selves and those of our physical, corporeal existence initially seem obvious. The corporeally embodied self is often viewed as a natural state of being, with our perspective, our consciousness, encapsulated within the shell of a physical body. On the other hand, virtual personas appear to be something that we build and craft, something that we create 'out there' in the world. However, these distinctions are not as definite as they first appear. This is because, as this text has asserted, we perform our corporeal selves and our virtual selves not as disembodied entities but as extensions of ourselves through the screen. Ultimately what is valued in the corporeal world is reflected onto the virtual, so that the search for difference and the potential for re-organised conditions of existence are doubtful. Even in the most utopian or subversive or transgressive spaces within platforms such as SL, there is a reaffirmation of the dominant paradigms of power and agency. Difference and the performance of another, more desired, self remain within the sphere of the established conditions of existence, most often reaffirming those of the mass-dominant culture which, in consideration of SL, is that of capitalist Western society. What has been particularly revealing when investigating SL for this book has been that the virtual can be so informative about the performative aspects of being. In highlighting our 'real' existence to be as much of a performance as our representations through the screen, we have an ability to reflect on nature and essentialism as phenomena that are created and informed by history and culture.

This text has shown that there is a high degree of similarity between screen culture and corporeal life which supports much of the previous research in the field of embodiment and screen practice. The analysis of SL within all chapters here has shown the replication of spatial conditions from the corporeal onto the virtual environment. Despite the ability to conceive space unbound by physicality, there is a very limited display of difference in the spaces. Moreover, the cultural practices of the corporeal have very definitely been reinscribed onto the spaces, with the majority of behavioural and representative practices very much like those of the contemporary Western world. This is particularly evident in the gendered and sexual representations and practices of the avatars that visit the spaces. The cyber-bodies of the avatars mirror those of the most desired bodies, creating an environment of bodily perfection. This is unsurprising given that the world of SL is highly predicated on CL, both in terms of the economic environment (being based on the late-capitalist model), as well as in the ingrained cultural practices of social and cultural ideologies that form the basis of envisaging worlds.

Whilst there are abundant similarities between the corporeal and virtual, there are also distinct differences in the way that the SL spaces are conceived, perceived and lived. This is principally because the environment is used mainly for pleasure and so the hedonistic elements of culture are brought to the fore. Although there are a number of SL residents who utilise the platform for work and business enterprises, the environment is primarily based on the notion of escapism. This is particularly true of the four regions of Apollo, Zeus, Eden and Greek Gold, which are framed as regions of fantasy and escape, suggestive of corporeal tourist spaces. The pleasure periphery of the virtual space is signified by the visual representation of the Mediterranean island and is extended to the spaces' shopping precincts and night-clubs, which work through the bodily beautiful avatars to represent idyllic and highly sexualised zones of pleasure and hedonism. It is for this reason that the gaze is evoked as a mechanism of defining the self and others. Power operates through the gaze, and ultimately operates very differently within the four spaces due to the respective differences seen in gender performance.

The medium of the screen also has a large effect on the way the environment is experienced. Representation, signification and the primacy of the gaze have substantial importance within the world of SL, so that the spatial characteristics of a space can be read as having specific meanings that are, in turn, specific to visual culture. Connotations of the classical world are written through each region, possibly being read as a part of the greater cultural imaginary that permeates mass culture. The multifarious nature of signification lends visual representation various points of reference and departure, yet there are often major ideological narratives that are written into, and derived from, popular texts. SL regions can be read as evoking popular tropes of signification as well as the iteration of Western symbolism, meaning that the classical world can be read as having a specific ideological meaning associated with popular history and culture. Meaning in this context is bounded by the gendered body in reference to the muscular beauty of the male form, and in the specific meanings of tolerance associated specifically with male same-sex sexual practice (as characterised by ancient Athens), and to a lesser extent the meanings attached to female space, physicality and sexuality (as associated with ancient Sparta and Biblical narratives).

Associated with the fantasy of the classical world and tourism is that of the potentially subversive. As spaces that are utilised for escape and desire, there is an assumption that many SL regions will be open and tolerant to diversity and difference. Many are regions that are indexical of subversion; however, the extent of this subversion does not necessarily extend to the representation of radical subjectivities. Subversion is signified throughout spaces that provide areas representative of hedonism, such as nightclubs and areas adorned with sexual imagery (Zeus), as well as spaces that provide alternative expression as commonplace. Radical subjectivities, however, were found primarily on Greek Gold through the disruption of normative notions of gender and sexuality. Other sites include the presence of 'furries' on Apollo, as well as some identities on Zeus who were engaging in practices that arguably define their sexuality and their interactions/relationships as outside of the normative. The signifying practices of these subjectivities, and what produces the meaning of the avatars and spaces, are expressed through the representational practices, particularly in reference to normativity or conformity. The representational practices on Zeus, whilst often very normative in terms of masculinity, also signify subversion through a highly overt sexuality. Within Greek Gold the avatars themselves signify subversion by their reluctance to fall easily within the corporeal identification of binary gender. Both Eden and Apollo, however, maintain normativity throughout spatial practice as well as avatar signification.

This performance of gender informs the assumption of gender and sexuality as necessary for the performance of identity. Ultimately, this contention is a critique of the assertion that in virtual spaces, and as part of a utopian technological future, gender will be made redundant as fluidity of identity becomes the norm. This possibility would see the dissolution of gendered categories and the undermining of essentialist positioning of gender and sexuality. As shown on the island of Greek Gold, there remains a desire to enact gender as a sexual performance and as a performance of identity. Gender performance within the space, rather than the dissolution of gender positions, is demonstrated as fluidity across gender performance. The spectrum of femininity to masculinity is displayed that reifies the reductive appropriation of norms and behaviours as untenable. By removing masculinity from the male, androgynous, butch and non-feminine identities can be viewed as having an enduring desire to perform, embody and identify with gender outside of the rigid binary relating to biological sex. The desired subject position of the corporeal masculine female is not a male self, but an idealised representation of her-self as a reinscribed female. Gender and sexuality then are not essential in terms of biological categories, but are necessary for the appropriation of identity and desire. Reification occurs through the interplay of object–subject positions which both interrupt and counteract the polemical understanding of gendered positions, sexual conformity and their associated normative behaviours. Unlike Eden, which re-presents standardised representations of the female body, Greek Gold shows a diversity of the female form. Historically conceived as a pathological form of inversion, the masculine female subject is often feared due to its disruption of normative binary gender and simultaneous disruption of the power positions afforded to the male and female subject. In a similar way to Zeus's positioning as subversive, so too are spaces like Greek Gold framed as all-too-different and divergent from mass culture and acceptance.

Limitations of this work and future considerations

As a predominant media of our time, screen technologies have had, and are having, a profound effect on how we communicate, interact and view ourselves. Our representations through the screen are, for many, a large part of our identity/identities, and we are able to play and manipulate these identities very easily through the representative power of the text. The major focus of this work has been to explore some of the suppositions about identity, specifically those of gender and sexuality. The use of technology, and specifically the SL platform, comes with a certain set of assumptions about the users and about culture in general. The viewpoint of this analysis is based in contemporary Western culture, specifically that of late-capitalist culture and more specifically the North Atlantic, in that North Americans and Europeans dominate usage of the SL platform. Politically and culturally, the US context dominates because SL is based on the free-market, deregulated economic system. It is also highly neo-liberal in terms of the allowance of behaviours and practices in view of economic agency. This Eurocentricism omits the experiences of a large portion of the world, and so deductively can only be seen to apply to the Western context. Furthermore, there is the high probability that platforms such as SL are utilised by those who are computer-literate, who have the capacity to purchase hardware that can run the SL software, and who have access to a high-speed internet connection. These factors position this work as a specifically middle-class project, which thus needs to be understood in the context of individuals who already possess a certain degree of education and agency in terms of identity expression. It would be a very interesting prospect to position this analysis outside of the Western, middle-class context and to investigate non-Western regions in terms of gender and sexuality.

It is also important to note here that this project, and its analysis, has been predominantly focused on white culture. During the research, there were very few avatars in the studied regions that displayed ethnic or non-white characteristics. As Lisa Nakamura (2008) contends, there is a very obvious racial framing of the internet and cyber-culture that privileges whiteness. In terms of visual culture, there are still advantages to being seen as white within virtual space as whiteness dominates as both a default avatar selection and in the selections available for purchase in the SL marketplace. Whilst outside the scope of this analysis, it would be of great benefit to analyse how non-white avatars are received within virtual spaces, and even to extend this investigation into regions of non-white gender and sexual diversity. It is often the case that non-white individuals feel doubly persecuted as non-heterosexual identities, both in the wider community and then again within LGBTI regions. It would be interesting to know the embodiment patterns of non-white individuals wanting to explore diverse regions in terms of maintaining or foregoing their racial identity.

It appears as though the possibility and desire to venture into virtual worlds will increase in the short- to mid-term future. The Second Life platform is looking to relaunch with an upgrade to Second Life called High Fidelity which is expected to vastly improve on aspects of real-time interaction that often hindered interactions in SL. Although the future success of the Second Life platform and its successor are uncertain given static SL numbers in recent years, there are a number of 3DVE platforms that have launched in the last few years, implying that there is a sizeable global market willing to sign up and experience these new worlds. Many of these platforms are marketed to specific communities, such as the teenage youth market (Club Cooee and vSide), adults (NuVera Online), those seeking mature content (Red Light Center and 3D Sex), and even specific ethnic groups (Gojiyo, India's first virtual world). Furthermore, the technology for producing and simulating environments within 3DVEs is increasing, along with the ability to more closely replicate aspects of embodiment that increase the feelings of presence and immersion, such as the embedding of 3D cameras in laptops that can detect movement and emotion. Facebook's acquisition of Oculus Rift (a virtual reality head-mounted display system) for US$2 billion points to there being some confidence in building upon the moderate success of virtual worlds to date.

As we move towards an even greater degree of technological convergence and so spend more time using visual interfaces, how we choose to represent ourselves becomes an even more pervasive characteristic of our identity. This representational choice is not something that has been readily acknowledged in terms of our corporeal lives, so the ability to actively choose visual characteristics as a modality of performance is relatively novel. It both opens up a whole new world of fluidity that allows for exploration and play, and reflects our corporeal identities in that we can consider the performativity of our bodies and our identities which are usually conceived of as natural. It is this dislodging of naturally ascribed characteristics that is the most important aspect of this work, and is one that is expected to become even more pervasive with the growing capabilities of screen technologies.

Notes

Introduction

1. User statistics for SL and its regions are difficult to come by, and, in the case of SL as a platform, need to be understood as self-reported numbers as part of a highly commodified environment. Region popularity is premised on the appearance of the site in SL's 'Destination Guide' (which may or may not be based on some financial arrangement with Linden Lab), as well as other, mainly subjective, factors that include external websites and the busyness of the site during the research itself.
2. On 26 June 2011, The Lost Gardens of Apollo was closed down by its creator Dane Zander, due to issues of island maintenance.

1 Embodiment, Virtual Experience and the Body: Possibilities for Subversion?

1. Of the ten default avatars available upon signing up to Second Life in 2011, only two were non-white. It is possible for users to change skin colour and other features within SL by buying accessories in the SL Marketplace. However, the marketplace is dominated by white avatars and add-ons of Caucasian features.

4 Masculinity, Mass Consumerism and Subversive Sex

1. The *Human Rights Watch* has various publications about global discrimination against LGBT individuals (LGBT Rights 2012). The statistics included as part of this study are specifically US figures to highlight the high incidence of anti-homosexual crimes in a nation that is also highly represented within SL user statistics.

References

Alexander, S. 2003. Stylish Hard Bodies: Branded Masculinity in Men's Health Magazine. *Sociological Perspectives,* 46, pp. 535–554.

Altman, D. 1996. Rupture or Continuity? The Internationalization of Gay Identities. *Social Text* 48, 77–94.

Alvesson, M. & Skoldberg, K. 2009. *Reflexive methodology: New vistas for qualitative research,* London, Great Britain, Sage.

Andersen, R. & Fetner, T. 2008. Economic Inequality and Intolerance: Attitudes Toward Homosexuality in 35 Democracies. *American Journal of Political Science,* 52, 942–958.

Ashdene, T. 2011. *Zeus Gay Club* [Online]. Available: http://zeusgayclub.com/?page_id=795 [Accessed 5 March 2012].

Avila-Saavedra, G. 2009. Nothing Queer About Queer Television: Televized Construction of Gay Masculinities. *Media, Culture & Society,* 31, 5–21.

Barthes, R. 2000. *Mythologies,* London, Great Britain, Vintage Books.

Baudrillard, J. 1983. *Simulations,* New York, NY, Semiotext(e).

Baudrillard, J. 1988. *The ecstasy of communication,* New York, NY, Semiotext(e).

Baudrillard, J. 2003. *Passwords,* London, Great Britain, Verso.

Bell, C. & Lyall, J. 2002. *The accelerated sublime: Landscape, tourism, and identity,* Westport, CT, Praeger.

Berger, J. 1972. *Ways of seeing,* London: Great Britain, BBC.

Bérubé, A. 2008. The History of Gay Bathhouses. *Journal of Homosexuality,* 44, 33–53.

Bhabha, H. K. 1998. The White Stuff. *Artforum,* 36(9), 21–22, 24.

Binnie, J. 1995. Trading places: Consumption, sexuality and the production of queer space. *In:* Bell, D. & Valentine, G. (eds.) *Mapping Desire: Geographies of Sexualities.* London: Routledge, pp. 145–160.

Binnie & Skeggs, B. 2004. Cosmopolitan Knowledge and the Production and Consumption of Sexualized Space: Manchester's Gay Village. *The Sociological Review,* 52, 39–61.

Biocca, F. (2006). The Cyborg's Dilemma: Progressive Embodiment in Virtual Environments. *Journal of Computer-Mediated Communication.* 3(2). Available: http://onlinelibrary.wiley.com/doi/10.1111/j.1083-6101.1997.tb00070.x/full

Bittarello, M. B. 2008. Another Time, Another Space: Virtual Worlds, Myths and Imagination. *Journal of Virtual Worlds Research,* 1.

Boellstorff, T. 2008. *Coming of age in second life,* New Jersey, NY, Princeton University Press.

Bordo, S. 1997. The body and the reproduction of femininity. *In:* Conboy, K., Medina, N. & Stanbury, S. (eds.) *Writing on the Body.* New York, NY: Colombia University Press.

Brown, D. 1996. Genuine fakes. *In:* Selwyn, T. (ed.) *The Tourist Image:Myths and Myth Making in Tourism.* Chichester, Great Britain: John Wiley and Sons.

Browne, K. & Bakshi, L. 2011. We are Here to Party? Lesbian, Gay, Bisexual and Trans Leisurescapes Beyond Commercial Gay Scenes. *Leisure Studies,* 30, 179–196.

Brownmiller, S. 1984. *Femininity,* New York, NY, Ballantine Books.

Brubaker, S. & Johnson, J. 2008. 'Pack a more powerful punch' and 'lay the pipe': Erectile Enhancement Discourse as a Body Project for Masculinity. *Journal of Gender Studies,* 17, 131–146.

Bryant, R. 2001. What Kind of Space is cyberspace? *Minerva – An Internet Journal of Philosophy,* 138–155.

Buchbinder, D. 1998. *Performance anxieties,* St Leonards, Australia, Allen & Unwin.

Butler, J. 1993. *Bodies that matter,* New York, NY, Routledge.

Butler, J. 1999. *Gender trouble,* New York, NY, Routledge.

Butler, J. 2004. *Undoing gender,* New York, NY, Routledge.

Campbell, J. E. 2004. *Getting it on online: Cyberspace, gay male sexuality, and embodied identity,* New York, NY, Harrington Park Press.

Card, C. 2006. The L Word and the F Word. *Hypatia,* 21, 223–229.

Cartledge, P. 2004. What Have the Spartans Done for us?: Sparta's Contribution to Western Civilization. *Greece and Rome,* 51, 164–179.

Casey 2001. Between Geography and Philosophy: What Does it Mean to be in the Place-World? *Annals of the Association of American Geographers,* 91, 683–693.

Casey, M. 2004. De-Dyking Queer Space(s): Heterosexual Female Visibility in Gay and Lesbian Spaces. *Sexualities,* 7, 446–461.

Castronova, E. 2005. *Synthetic worlds: The business and culture of online games,* Chicago, IL, The University of Chicago Press.

Clark, J. 2011. Second chances: Depictions of the natural world in Second Life. *In:* Ensslin, A. & Muse, E. (eds.) *Creating Second Lives: Community, Identity and Spatiality as Constructions of the Virtual.* New York, NY: Routledge.

Cohen, E. 1979. A Phenomenology of Tourist Experiences. *Sociology,* 13, 179–201.

Connell, R. W. 2005. Change Among the Gatekeepers: Men, Masculinities, and Gender Equality in the Global Arena. *Signs,* 30, 1801–1825.

Corsianos, M. 2007. Mainstream Pornography and 'women': Questioning Sexual Agency. *Critical Sociology,* 33, 863–885.

Cover, R. 2006. Material/Queer Theory: Performativity, Subjectivity, and Affinity-Based Struggles in the Culture of Late Capitalism. *Rethinking Marxism,* 16, 293–310.

Cover, R. 2010. Object(ive)s of Desire: Romantic Coupledom Versus Promiscuity, Subjectivity and Sexual Identity. *Continuum: Journal of Media & Cultural Studies,* 24, 251–263.

Coy, M. 2009. Milkshakes, Lady Lumps and Growing up to Want Boobies: How the Sexualisation of Popular Culture Limits Girls' Horizons. *Child Abuse Review* 18, 372–383.

Coyne, R. 1998. Cyberspace and Heidegger's Pragmatics. *Information Technology and People,* 11, 338–350.

Crawshaw, C. & Urry, J. 1997. Tourism and the photographic eye. *In:* Rojek, C. & Urry, J. (eds.) *Touring Cultures: Transformations of Travel and Theory.* London, Great Britain: Routledge.

Creed, B. 1995. Lesbian bodies: Tribades, tomboys and tarts. *In:* Grosz, E. & Probyn, E. (eds.) *Sexy Bodies: The Strange Carnalities of Feminism.* London: Great Britain: Routledge.

Csordas, T. 1994. The body as representation and being-in-the-world. *In:* Csordas, T. (ed.) *Embodiment and Experience.* Cambridge, Great Britain: Cambridge University Press.

De Certeau, M. 1984. Walking the city. *Practices of everyday life.* Berkeley: University of California Press.

Dean, T. 2009. *Unlimited intimacy: Reflections on the subculture of barebacking,* Chicago, IL, University of Chicago Press.

Dell, K. 2008. How second life affects real life. *Time.*

Dethridge, L. 2009. The Magic Realism of a Virtual Second Life. *Literature & Aesthetics,* 19.

Dholakia, R. 2007. Gender and IT in the Household: Evolving Patterns of Internet Use in the United States. *The Information Society,* 22, 231–240.

Diamond, L. 2005. 'I'm straight, but I kissed a girl': The Trouble with American Media Representations of Female-Female Sexuality. *Feminism & Psychology,* 15.

Diken, B. & Laustsen, C. 2004. Sea, Sun, Sex and the Discontents of Pleasure. *Tourist Studies,* 4, 99–114.

Dorfler, M. 2010. *Use the Age Verification System Already in Place to do a Voluntary Gender Verification to Protect Against the Fraudulent use of Female Avatars by RL Males* [Online]. Available: https://jira.secondlife.com/browse/SVC-4180 [Accessed 24 February 2011].

Dotson, M., Hyatt, E. & Thompson, L. 2009. Sexual Orientation and Gender Effects of Exposure to Gay- and Lesbian-Themed Fashion Advertisements. *Journal of Fashion Marketing and Management,* 13, 431–447.

Dover, K. J. (2002). Classical Greek Attitudes to Sexual Behavior. *In*: McClure, L. (ed.) *Sexuality and gender in the Classical world.* Oxford: Blackwell, pp. 19–34.

Dowsett, G., Williams, H., Ventuneac, A. & Carballo-Dieguez, A. 2008. 'Taking it like a man': Masculinity and Barebacking Online. *Sexualities,* 11, 121–141.

Doyle, D. 2011. The body of the avatar. *In:* Ensslin, A. & Muse, E. (eds.) *Creating Second Lives.* New York, NY: Routledge.

Ducheneaut, N., Wen, M.-H., Yee, N. & Wadley, G. 2009. *Body and mind: A study of avatar representation in three virtual worlds. CHI 2009.* Boston: ACM.

Duncan, J. & Duncan, N. 1992. Ideology and bliss: Roland barthes and the secret histories of landscape. *In:* Barnes, T. & Duncan, J. (eds.) *Writing worlds: Discourse and Metaphor in the Representation of Landscape.* London, Great Britain: Routledge.

Dyer, R. 2002. *The culture of queens,* London, Great Britain, Routledge.

Easthope, A. 1992. *What a man's gotta do,* New York, NY, Routledge.

Edwards, T. 1994. *Erotics and politics: Gay male sexuality, masculinity and feminism,* London, Great Britain, Routledge.

Edwards, T. 2006. *Cultures of masculinity,* Milton Park, Great Britain, Routledge.

Ensslin, A. 2011. Avatar needs and the remediation of architecture in second life. *In:* Ensslin, A. & Muse, E. (eds.) *Creating Second Lives.* New York, NY: Routledge.

Eos 2008. [Online]. Available: http://www.theoi.com/Titan/Eos.html [Accessed 20 March 2010].

Europeans predominate as Second Life users: study. 2007. Available: http://www.reuters.com/article/2007/05/04/us-secondlife-residents-idUSL0445387420070504 [Accessed 1 June 2012].

Evans 1983. *Woman in the bible,* Carlisle, Great Britain, Paternoster Press.

Evans & Gamman, L. 1995. The gaze revisited, or reviewing queer viewing. *In:* Burston, P. & Richardson, C. (eds.) *A Queer Romance: Lesbians, Gay Men and Popular Culture.* London: Great Britain: Routledge.

Eves, A. 2004. Queer Theory, Butch/Femme Identities and Lesbian Space. *Sexualities*, 7, 480–496.

Fine, C. 2010. *Delusions of gender: The real science behind sex differences*, London: Great Britain, Icon Books.

Finlayson, I. 2012. The Human Condition: Johnny be Good *The Independent* [Online]. Available: http://www.independent.co.uk/life-style/the-human-condition-johnny-be-good-1166398.html [Accessed 20 September 2012].

Fischler, S. 1998. Imperial cult: Engendering the cosmos. *In:* Foxhall, L. & Salmon, J. (eds.) *When Men Were Men*. London, Great Britain: Routledge.

Fox, M. 1998. The constrained man. *In:* Foxhall, L. & Salmon, J. (eds.) *Thinking Men*. Routledge: London, Great Britain.

Fuchs, C. 1995. 'Death is irrelevant': Cyborgs, reproduction, and the future of male hysteria. *In:* Hables, C. (ed.) *The Cyborg Handbook*. New York, NY: Routledge.

Gay destinations 2012a. [Online]. Second Life. Available: http://search.secondlife.com/?query_term=gay+&collection_chosen=destinations&isLeftColOpen=true&sort=dg_order_asc&category=0 [Accessed 5 March 2012].

Gibbs, R. 2006. *Embodiment and cognitive science*, New York, NY, Cambridge University Press.

Giblett, R. 2011. *People and places of nature and culture*, Bristol, Great Britain, Intellect.

Gill, R. 2008. Empowerment/Sexism: Figuring Female Sexual Agency in Contemporary Advertising. *Feminism & Psychology*, 18, 35–60.

Gill, R., Henwood, K. & Mclean, C. 2005. Body Projects and the Regulation of Normative Masculinity. *Body and Society*, 11, 37–62.

Gorman-Murray, A. & Waitt, G. (2009). Queer-Friendly Neighbourhoods: Interrogating Social Cohesion across Sexual Difference in Two Australia Neighbourhoods. *Environment and Planning A*, 41(12), 2855–2873.

Gosine, A. 2007. Brown to blonde at gay.com: Passing white in queer cyberspace. *In:* O'riordon, K. & Phillips, D. J. (eds.) *Queer Online: Media Technology and Sexuality*. New York, NY: Peter Lang.

Gotta, G., Green, R.-J., Rothblum, E., Solomon, S., Balsam, K. & Schwartz, P. 2011. Heterosexual, Lesbian, and Gay Male Relationships: A Comparison of Couples in 1975 and 2000. *Family Process*, 50, 353–376.

Gottdiener, M. 1995. *Postmodern semiotics: Material culture and the forms of postmodern life*, Oxford, Great Britain, Blackwell.

Graburn, N. 1983. The Anthropology of Tourism. *Annals of Tourism Research*, 10, 9–33.

Gray, P. 1970. *International travel – International trade*, Lexington, MA, Heath Lexington Books.

Greek Gold Lesbian Resort 2011a. [Online]. Available: http://world.secondlife.com/group/49a9e0dd-ed26-5370-1780-0da809cdb35c [Accessed 1 March 2011].

Grosz, E. 1995. *Space, time and perversion*, London, Great Britain, Routledge.

Grosz, E. 2001. *Architecture from the outside*, Cambridge, NY, MIT Press.

Guyer, P. 2006. *Kant*, New York, NY, Routledge.

Hafkin, N. & Huyer, S. 2007. Women and Gender in ICT Statistics and Indicators for Development. *Information Technologies and International Development*, 4, 25–41.

Halberstam, J. 1998. *Female masculinity,* London, Great Britain, Duke University Press.

Halkitis, P., Parsons, J. & Wilton, L. 2003. Barebacking Among Gay and Bisexual Men in New York City: Explanations for the Emergence of Intentional Unsafe Behavior. *Archives of Sexual Behaviour,* 32, 351–357.

Hall, S. 1996. Cultural studies and its theoretical legacies. *In:* Morley, D. & Chen, K.-H. (eds.) *Critical dialogues in cultural studies.* London, Great Britain: Routledge.

Hallett, J. 1979. Sappho and her Social Context: Sense and Sensuality. *Signs,* 4, 447–464.

Haraway, D. 1991. *Simians, cyborgs, and women: The reinvention of nature,* New York, NY, Routledge.

Harkin, M. 1995. Modernist Anthropology and Tourism of the Authentic. *Annals of Tourism Research,* 22.

Harris, K. 1984. *Sex, ideology and religion: The representation of women in the Bible,* Brighton, Great Britain, Wheatsheaf Books.

Harris & Platzner, G. 1998. *Classical mythology: Images and insights,* London, Great Britain, Mayfield.

Hemmings, C. 1998. Waiting for no man: Bisexual femme subjectivity and cultural repudiation. *In:* Munt, S. (ed.) *Butch/Femme: Inside Lesbian Gender.* London, Great Britain: Cassell.

Hera Emms. 2011a. Welcome Note.

Hera Emms 2011b. [Online]. Available: http://world.secondlife.com/resident/ccbeb623-9190-4cdb-95db-19e28990e8f2 [Accessed 1 March 2011].

Hill, S. 2006. *Winged Beings in Ancient Greek art* [Online]. Available: http://arthistory.suite101.com/article.cfm/winged_beings_in_ancient_greek_art [Accessed 23 March 2010].

Hillman, W. 2007. Revisiting the concept of (objective) authenticity. *TASA.* Auckland.

Hubbard, T. 2009. The Paradox of 'Natural' Heterosexuality with 'Unnatural' Women. *Classical World,* 102, 249–258.

Inness, S. 1999a. *Tough girls,* Philadelphia, PA, Penn.

Inness, S. 1999b. *Tough girls: Women warriors and wonder women in popular culture,* Philadelphia, PA, University of Pennsylvania Press.

Jackson, S. 2009. 'Hot lesbians': Young People's Talk About Representations of Lebianism. *Sexualities,* 12, 199–224.

Jagger, G. 2008. *Judith Butler: Sexual politics, social change and the power of the performative,* New York, Routledge.

James AU, W. 2006. *New World SLURL: The Lost Gardens of Apollo* [Online]. Available: http://nwn.blogs.com/nwn/2006/03/new_world_slurl_9.html [Accessed 20 Feb 2010].

Justice, U. D. O. 2009. *Hate Crime Statistics, 2009* [Online]. Available: http://www2.fbi.gov/ucr/hc2009/victims.html [Accessed 6 September 2011].

Kennedy, G. 1991. Hemingway's Gender Trouble. *American Literature,* 63, 187–207.

Kimmel, M. 2001. Masculinity as homophobia: Fear, shame and silence in the construction of gender identity. *In:* Whitehead, S. & Barrett, F. (eds.) *The Masculinities Reader.* Cambridge, Great Britain: Polity Press.

King, H. 1993. Bound to bleed: Artemis and Greek women. *In:* Cameron, A. & Kuhrt, A. (eds.) *Images of Women in Antiquity.* Milton Park, Great Britain: Routledge.

Kroker, A. 1996. Virtual capitalism. *In:* Aronowitz, S., Martinsons, B. & Menser, M. (eds.) *Technoscience and Cyberculture*. New York, NY: Routledge.

Leder, D. 1990. *The Absent Body,* Chicago, University of Chicago Press.

Lefebvre, H. 1991. *The production of space,* Oxford, Great Britain, Blackwell.

Lefkowitz, M. 1986. *Women in Greek myth,* London, Great Britain, Gerald Duckworth and Co.

Lefkowitz, M. 2002. 'Predatory' Goddesses. *Hesperia,* 71, 325–344.

Lett, J. 1983. Ludic and Liminoid Aspects of Charter Yacht Tourism in the Britsih Virgin Islands. *Annals of Tourism Research,* 10, 35–56.

Levy, A. 2005. *Female chauvinist pigs: Women and the rise of raunch culture,* Melbourne, Australia, Schwartz Publishing.

Light, B., Fletcher, G. & Adam, A. 2008. Gay men, Gaydar and the Commodification of Difference. *Information Technology and People,* 21, 300–314.

Linden. 2011a. *Maturity Ratings* [Online]. Available: http://community.secondlife.com/t5/English-Knowledge-Base/Maturity-ratings/ta-p/700119#Maturity_ratings_overview [Accessed 27 April 2011].

Linden. 2011b. *Q1 2011 Linden Dollar Economy Metrics up, Users and Usage Unchanged* [Online]. Available: http://community.secondlife.com/t5/Featured-News/Q1-2011-Linden-Dollar-Economy-Metrics-Up-Users-and-Usage/ba-p/856693 [Accessed 5 January 2012].

Linden Lab Official: Clarification of policy disallowing ageplay 2011c. [Online]. Available: http://wiki.secondlife.com/wiki/Linden_Lab_Official:Clarification_of_policy_disallowing_ageplay [Accessed 13 June 2011].

Maccannell, D. 1999. *The tourist: A new theory of the leisure class,* Berkeley, CA, University of California Press.

Mackinnon, K. 1997. Uneasy pleasures: The male as erotic object. London, Great Britain: Cygnus Arts.

Marin, L. 1993. Frontiers of Utopia: Past and Present. *Crtitical Inquiry,* 19, 397–420.

Markwick, M. 2001. Postcards from Malta: Image, Consumption, Context. *Annals of Tourism Research,* 28, 417–438.

Marzullo, M. 2011. Through a Glass, Darkly: US Marriage Discourse and Neoliberalism. *Journal of Homosexuality,* 58, 758–774.

Mclaughlin, J., Casey, M. & Richardson, D. 2006. At the intersections of feminist and queer debates. *In:* Mclaughlin, J., Casey, M. & Richardson, D. (eds.) *Intersections Between Feminist and Queer Theory*. New York, NY: Palgrave Macmillan.

Mercer, K. 1991. Looking for Trouble. *Transition,* 51, 184–197.

Merleau-Ponty, M. 1964. *The primacy of perception,* Evanston, IL, Northwestern University Press.

Messinger, P., Stroulia, E., Lyons, K., Bone, M., Niu, R., Smirnov, K. & Perelgut, S. 2009. Virtual Worlds — Past, Present, and Future: New Directions in Social Computing. *Decision Support Systems,* 47.

Metzl, J. 2004. From Scopophilia to *Survivor*: A Brief History of Voyeurism. *Textual Practice,* 18, 415–434.

Moore, B. & Isen, A. 1990. Affect and social behaviour. *In:* MOORE, B. & ISEN, A. (eds.) *Affect and Social Behaviour*. Cambridge, Great Britain: Cambridge University Press.

Montserrat, D. 1998. Experiencing the male body in Roman Egypt. *In:* Foxhall, L. & Salmon, J. (eds.) *When Men Were Men: Masculinity, Power and Identity in Classical Antiquity*. New York, NY: Routledge.

Mosco, V. 2004. *The digital sublime: Myth, power, and cyberspace,* Cambridge, MA, MIT Press.

Mowlabocus, S. 2008. Revisiting Old Haunts Through New Technologies: Public (homo)Sexual Cultures in Cyberspace. *International Journal of Cultural Studies,* 11, 419–438.

Moylan, T. 1986. *Demand the impossible: Science fiction and the utopian imagination,* New York, NY, Methuen.

Mulvey, L. 1975. Visual Pleasure and Narrative Cinema. *Screen,* 16.

Muse, E. 2011. The event of space: Defining place in a virtual landscape. *In:* Ensslin, A. & Muse, E. (eds.) *Creating Second Lives.* New York, NY: Routledge.

Nakamura, L. 2008. *Digitizing race:Visual cultures of the internet,* Minneapolis, MN, University of Minnesota Press.

Nestle, J. 1992. The femme question. *In:* Nestle, J. (ed.) *The Persisten Desire: A Femme-Butch Reader.* Boston: MA: Alyson Publications.

Nowak, K. & Rauh, C. 2008. Choose Your 'buddy icon' Carefully: The Influence of Avatar Androgyny, Anthropomorphism and Credibility in Online Interactions. *Computers in Human Behaviour,* 24, 1473–1493.

Nye, D. 1994. *American technological sublime,* Cambridge, MA, MIT Press.

Overlord, R. 2007. *The Lost Gardens of Apollo* [Online]. Available: http://www.secondlifetraveler.com/category/romantic-spots/lost-gardens-of-apollo [Accessed 16 March 2009].

Pearce, P. & Moscardo, G. 1986. The Concept of Authenticity in Tourist Experiences. *Journal of Sociology,* 22.

Peters, B. M. 2003. Qu(e)erying Comic Book Culture and Representations of Sexuality in Wonder Woman. *CLCWeb: Comparative Literature and Culture,* 5.

Peterson, G. T. 2011. Clubbing Masculinities: Gender Shifts in Gay Men's Dance Floor Choreographies. *Journal of Homosexuality,* 58, 608–625.

Prigge, W. 2008. Reading the urban revolutiuon: Space and representation. *In:* Goonewardena, K. (ed.) *Space, Difference, Everyday Life.* New York, NY: Routledge.

Pritchard, A. & Morgan, N. 2000. Privileging the Male Gaze: Gendered Tourism Landscapes. *Annals of Tourism Research,* 27, 884–905.

Pritchard, A., Morgan, N. & Sedgley, D. 2007. In search of lesbian space? The experience of Manchester's gay village. *In:* Pritchard, A., Morgan, N., Ateljevic, I. & Harris, C. (eds.) *Tourism and Gender: Embodiment, Sensuality and Experience.* Oxfordshire, Great Briatin: CABI.

Puar, J. 2003. A Transnational Feminist Critique of Queer Tourism. *Antipode,* 34, 935–946.

Pullen, C. 2010a. *LGBT identity and online new media.* New York, NY: Routledge.

Pullen, C. 2010b. The murder of Lawrence king and the LGBT online stimulations of narrative copresence. *LGBT identity and online new media.* New York, NY: Routledge.

Queen, C. & Comella, L. 2008. The Necessary Revolution: Sex-Positive Feminism in the Post-Barnard Era. *The Communication Review,* 11, 274–291.

Relph, E. 2001. Sense of place. *In:* Hanson, S. (ed.) *Ten Geographic Ideas that Changed the World.* New Jersey, NY: Rutgers State University.

Richardson, D. 2006. Bordering theory. *In:* Mclaughlin, J., Casey, M. & Richardson, D. (eds.) *Intersections Between Feminist and Queer Theory.* New York, NY: Plagrave Macmillan.

Robins, K. 1996. *Into the image: Culture and politics in the field of vision,* London, Great Britain, Routledge.

Robinson, P. 2005. *Queer wars,* Chicago, IL, The University of Chicago Press.

Rohlinger, D. 2002. Eroticizing Men: Cultural Influences on Advertising and Male Objectification. *Sex Roles,* 46.

Rojek, C. 1993. *Ways of escape: Modern transformations in leisure and travel,* Basingstoke, Great Britain, Macmillan Press.

Rojek, C. 1997. Indexing, dragging and the social construction of tourist sights. *In:* Rojek, C. & Urry, J. (eds.) *Touring Cultures.* London, Great Britain: Routledge.

Roof, J. 1998. 1970s lesbian feminism meets 1990s butch-femme. *In:* Munt, S. (ed.) *Butch/Femme: Inside Lesbian Gender.* London, Great Brritain: Cassell.

Rose, G. 1993. *Feminism and geography: The limits of geographical knowledge,* Cambridge, Great Britain, Polity Press.

Ryan, C. 2002. Memories of the beach. *In:* Ryan, C. (ed.) *The Tourist Experience.* London, Great Britain: Continuum.

Sabine666 Kuhn. 2012. *Sister of Eden* [Online]. Available: http://world.secondlife.com/group/2f1c6027-51d0-e2bd-72cd-c5f17f350e19 [Accessed 25 January 2012].

Sartre, J.-P. 1943. *Being and nothingness,* London, Great Britain, Routledge.

Second Life marketplace 2012b. [Online]. Available: https://marketplace.secondlife.com/products/search?search%5Bcategory_id%5D=4&search%5Bmaturity%5D%5B%5D=General&search%5Bmaturity%5D%5B%5D=Moderate&search%5Bmaturity%5D%5B%5D=Adult&search%5Bpage%5D=1&search%5Bper_page%5D=12 [Accessed 19 March 2012].

Second Life places 2009. [Online]. Available: http://world.secondlife.com/place/65f385f0-691c-5755-9ef0-15f3bb05c8b7 [Accessed 20 February 2010].

Selanniemi, T. 2003. On holiday in the liminoid playground: Place, time, and self in tourism. *In:* Bauer, T. G. & Mckercher, B. (eds.) *Sex and Tourism: Journeys of Romance, Love, and Lust.* London, Great Britain: The Halworth Hospitality Press.

Sender, K. 2003. Sex sells: Sex, Class, and Taste in Commercial Gay and Lesbian Media. *Journal of Lesbian and Gay Studies,* 9, 331–365.

Sender, K. 2004. *Business not politics: The making of the gay market,* New York, NY, Columbia University Press.

Sermon, P. & Gould, C. 2011. Liberate your avatar: The revolution will be socially networked. *In:* Ensslin, A. & Muse, E. (eds.) *Creating Second Lives.* New York, NY: Routledge.

Sharpley, R. 2004. Islands in the sun: Cyprus. *In:* Urry, J. & Sheller, M. (eds.) *Tourism Mobilities: Place to Play, Places in Play.* London, Great Britain: Routledge.

Shepherd, T. 2014. Second Life Grid Survey – Region Database [Online]. Linden Lab. [Accessed 8 Sept 2014].

Simpson, M. 1994. *Male impersonators: Men performing masculinity,* New York, NY, Routledge.

Sofoulis, Z. 2002. Cyberquake: Haraway's manifesto. *In:* Jähnert, G., Aleksander, K., & Rosenbusch, K. (eds) *Cyberfeminismus: feministische Visionen mit Netz und ohne Boden?* Berlin: Humboldt University, pp. 54–72. [shortened version published by permission of D Tofts, A Jonson & A Cavallaro 2003].

Soja, E. 1989. *Postmodern geographies: The reassertion of space in critical social theory,* New York, NY, Verso.

Stacey, J. 2004. Cruising to Familyland: Gay Hypergamy and Rainbow Kinship. *Current Sociology,* 52, 181–197.

Stewart, A. 1997. *Art, desire, and the body in ancient Greece,* New York, NY, Cambridge University Press.

Sunden, J. 2003. *Material virtualities,* New York, NY, Peter Lang.

Taylor, J. (2008). The Queerest of Queer: Sexuality, Politics and Music on the Brisbane Scene. *Continuum.*·22(5), pp. 651–665.

Tomsen, S. & Markwell, K. 2009. Violence, Cultural Display and the Suspension of Sexual Prejudice. *Sexuality & Culture,* 13.

Turner T. 1994. Bodies and anti-bodies: Flest and fetish in contemporary social theory. *In:* CSORDAS, T. (ed.) *Embodiment and Experience.* Cambridge, Great Britain: Cambridge University Press.

Turner, V. 1977. *Secular Ritual.* Assen: Van Gorcum.

Turner, L. & Ash, J. 1975. *The golden hordes,* London, Great Britain, Constable and Company.

Urry, J. 2002. *The tourist gaze,* London, Great Britain, Sage.

Urry, J. & Larsen, J. 2011. *The tourist gaze 3.0,* London, Gerat Britain, SAGE.

Valentine, G. 1993. Negotiating and Managing Multiple Sexual Identities: Lesbian Time-Space Strategies. *Transactions of the Institute of British Geographers,* 18, 237–248.

Valentine, G. 1996. (Re)Negotiating the heterosexual street: Lesbian productions of space. *In:* Duncan, N. (ed.) *BodySpace.* London: Great Britain: Routledge.

Valentine, G. & Skelton, T. 2003. Finding Oneself, Losing Oneself: The Lesbian and Gay 'scene' as a Paradoxical Space. *International Journal of Urban and Regional Research,* 27, 849–866.

Vander Valk, F. 2008. Identity, Power, and Representation in Virtual Environments. *MERLOT Journal of Online Learning and Teaching,* 4.

Varela, F., Thompson, E. & Rosch, E. 1991. *The embodied mind: Cognitive science and human experience,* Cambridge, MA, MIT Press.

Vicinus, M. 1992. 'They wonder to which sex I belong': The Hisorical Roots of the Modern Lesbian Identity. *Feminist Studies,* 18, 467–497.

Virilio, P. 1991. *The lost dimension,* New York, NY, Semiotext(e).

Virilio, P. 2000. *Polar inertia,* London, Great Britain, Sage.

Waitt, G. & Markwell, K. 2006. *Gay tourism: Culture and context,* New York, NY, The Haworth Hospitality Press.

Wajcman, J. 1994. Technological a/genders: Technology, culture and class. *In:* Lelia., G. & Guinery, R. (eds.) *Framing technology.* St Leonards, Australia: Allen & Unwin.

Warner, M. 1999. *The trouble with normal: Sex, politics and the ethics of queer,* Harvard University Press.

Warner, M. 2000. *The trouble with normal: Sex, politics and the ethics of queer life,* Cambridge, MA, Harvard University Press.

Wearing, S., Stevenson, D. & Young, T. 2010. *Tourist cultures: Identity, place and the traveller,* London, Great Britain, Sage.

Wertheim, M. 1999. *The pearly gates of cyberspace,* New York, NY, Random House.

Wertsch, J. 1985. *Vygotsky and the Social Formation of Mind,* Cambridge, Harvard University Press.

Whitley, J. 2001. *The archaeology of ancient Greece,* Cambridge, Great Britain, Cambridge University Press.

Williams, L. 1999. *Hard core: Power, pleasure, and the 'frenzy of the visible',* Berkeley, CA, University of California Press.

Williams, L. 2007. The pornography of consumption/The consumption of pornography. *In:* Herdt, G. & Howe, C. (eds.) *21st Century Sexualities: Contemporary Issues in Health, Education and Rights.* New York, NY: Routledge.

Woods, G. 1995. Fantasy islands: Popular topographies of marooned masculinity. *In:* Bell, D. & Valentine, G. (eds.) *Mapping Desire.* London, Great Britain: Routledge.

Yee, N. 2009. *Our Virtual Bodies, Ourselves?* [Online]. Available: http://www.nickyee.com/daedalus/archives/001613.php?page=1 [Accessed 20 June 2009].

Yee, N. & Bailenson, J. 2009. The proteus effect: Implications of transformed digital self-representation on online and offline behavior. Stanford University.

Zander, D. 2006. Welcome note.

Ziemke, T. Year. What's that thing called embodiment? *In:* Erlbaum, L. (ed.) 25th Annual Meeting of the Cognitive Science Society, 2003.

Zeus Gay Club 2012c. [Online]. Available: http://zeusgayclub.com/ [Accessed 23 May 2012].

Index